# Thermal Energy Waste Recovery Technologies and Systems

*Thermal Energy Waste Recovery Technologies and Systems* comprehensively covers thermal energy recovery technologies and systems. It considers thermal sources, working principles, products, application status, prospects, and challenges.

In an effort to achieve energy security, carbon neutralization, and sustainable development, this book discusses waste recovery from thermal energy technologies and systems from varying temperatures. This book features case studies of advanced multi-generation systems for different industrial applications.

This book is intended for senior undergraduate and graduate mechanical engineering students taking courses in thermal energy, energy systems, and renewable energy, as well as researchers studying thermal energy utilization, low-carbon technologies, thermal dynamic analysis, and energy system design.

# Thermal Energy Waste Recovery Technologies and Systems

Chen Weidong, Huang Zhifeng, and
Chua Kian Jon

CRC Press
Taylor & Francis Group
Boca Raton London New York

CRC Press is an imprint of the
Taylor & Francis Group, an **informa** business

Designed cover image: Shutterstock

First edition published 2024
by CRC Press
6000 Broken Sound Parkway NW, Suite 300, Boca Raton, FL 33487-2742

and by CRC Press
4 Park Square, Milton Park, Abingdon, Oxon, OX14 4RN

*CRC Press is an imprint of Taylor & Francis Group, LLC*

ISBN: 978-1-032-38070-4 (hbk)
ISBN: 978-1-032-38075-9 (pbk)
ISBN: 978-1-003-34338-7 (ebk)

DOI: 10.1201/9781003343387

Typeset in Times
by codeMantra

# Contents

# Preface

Due to existing issues related to global energy security and ecological environment degradation, waste thermal energy recovery research and development has gained traction over the last 20 years. It is imperative to develop and deploy waste thermal energy recovery technologies for several key reasons. The first reason is to achieve the goal of carbon neutrality. A significant amount of waste heat produced by various industrial activities has been consistently being discharged to the ambient environment every year, creating additional carbon footprint problems. Thus, the deployment of waste heat recovery technologies enables the reduction of greenhouse gas emissions. The second reason is to promote the economic well-being and energy-efficient performance of industrial activities. It is apparent that harnessing the recoverable waste energy improves energy efficiency and reduces operating costs in industrial activities. The third reason being that it is the common responsibility and moral obligation of a global citizen to reduce waste and employ available thermal energy to sustain day-to-day operations.

Waste thermal energy sources manifest in different forms and types (solid, liquid, gas) spanning different temperature ranges (160°C–1,800°C). The recoverable thermal energy can be harnessed from fuel-driven prime movers, biomass energy, fuel cell technologies, renewable heat energy, data centers, nuclear plants, and various other industrial activities. In addition, useful utilities, namely, electricity, cooling, heating, potable water, and dry and cool air, can be produced from a judiciously integrated thermal energy system that is fed from the harvested waste thermal heat. Energy storage technologies are sometimes required to store the waste thermal energy since the energy-consuming site may be located far away from the waste energy production source.

This monograph starts by providing a comprehensive analysis of various waste thermal energy recovery technologies based on the nature of their heat sources and products. One of the chapters conducts thermal dynamic analyses to facilitate an in-depth understanding of various waste heat-driven technologies. Another chapter focuses on multi-generation systems integrated with waste thermal energy recovery technologies to demonstrate the efficient use of waste energy to promote energy efficiency. Implementing a proper waste heat recovery technology is key toward designing a feasible multigeneration system with decent energy utilization efficiency. The remaining chapters of the monographs cover case studies of advanced multigeneration systems for different industrial applications. The last chapter highlights the current status, challenges, and prospects of employing waste thermal energy technologies, and designing of innovative thermal systems is also described and documented to inspire further research and applications.

**Chen Weidong**
**Huang Zhifeng**
**Chua Kian Jon**

# About the Authors

**Dr. Chen Weidong** is currently a research fellow from the Department of Mechanical Engineering, National University of Singapore (NUS). He completed his undergraduate coursework at Tongji University, during which time he received the first prize scholarship every year and won the Outstanding Graduates of Shanghai in 2018. As the recipient of the NUS research scholarship, he started his Ph.D. in Mechanical Engineering from NUS in 2018, under Prof. Kian Jon Chua's supervision. During his doctoral training, he focused on the experiment and simulation investigation of a multigeneration system to achieve improved energy performance. The multigeneration system is able to high-efficiently produce electricity, cooling, heating, potable water, dry and cool air by harnessing waste thermal energy. So far, he has published six research papers in the top *Journal of Applied Energy (APEN)*, *Energy Conversion and Management (ECM)*, and *Energy*. His specific research focus areas include CCHP, waste heat recovery, temperature cascade utilization system, adsorption chiller, absorption chiller, multi-objective optimization, optimal operating strategy analysis, and system simulation by employing Monte Carlo simulation and Simulink platform.

**Mr. Huang Zhifeng** is currently pursuing his Ph.D., as the recipient of the prestigious NUS Graduate scholarship, in the Department of Mechanical Engineering, National University of Singapore (NUS), under Prof. Kian Jon Chua's supervision. He received his Bachelor's and Master's degrees from SCUT, China, and worked as an engineer in South China Power Gird for 3 years. During his doctoral training, he designed a series of advanced systems to improve the energy utilization efficiency and recovery rate for LNG cold energy, as well as experiment on a cascading LNG cold energy recovery experimental setup in the lab. His research findings have been published in various top international journals, including *Applied Energy*, *Energy Conversion Management*, and *Energy*. Huang's specific research areas focus on gas turbines, co-generation systems, tri-generation systems, waste heat recovery, and waste cold energy recovery.

**Dr. Chua Kian Jon** is an associate professor with the Department of Mechanical Engineering, National University of Singapore. He has been conducting research on air-conditioning, refrigeration, and heat recovery systems since 1997. He has conducted both modeling and experimental works for specific thermal energy systems, including dehumidification, cooling, heat pumping, compact heat exchangers, and refined temperature/humidity control. He is highly skilled in designing; fabricating; commissioning; and testing many sustainable energy systems to provide for heating, cooling, and humidity control for both small- and large-scale applications.

Dr. Chua has published over 200 international peer-reviewed journal publications, six book chapters, and two recent monographs on advances in air conditioning and dehumidification technologies. He was highlighted to be among the top 2% of global energy scientists based on Elsevier's database, 1% of scientists in the world by the Universal Scientific Education and Research Network, and top 0.3% in the Stanford list of energy researchers. He has been elected to several fellowships including Fellow of Institute of Mechanical Engineering, Fellow of Royal Society, and Fellow of Energy Institute. His works have garnered more than 11,500 over citations with a current h-index of 56. Further, he owns more than 15 patents related to several innovative cooling and dehumidification systems. He is the principal investigator of several multi-million competitive research grants. Additionally, he has been awarded multiple local, regional, and international awards for his breakthrough research endeavors.

# 1 Sources of Recoverable Waste Thermal Energy

## 1.1 BYPRODUCTS FROM PRIME MOVERS

Prime movers consume primary energy to generate electricity, and a huge amount of recoverable thermal heat is produced and not fully harnessed. If this thermal heat is exhausted into the environment and not recovered, a large quantity of useful energy is wasted. Therefore, it is important to understand prime movers' working principles and their energy-related byproducts for improved waste heat utilization. In addition, the temperatures and types of dissipated heat have significant impacts on heat recovery performance. The following sections introduce the working principles, mathematical models, and application status of gas turbines, internal combustion engines, Stirling engines, steam turbines, and fuel cells.

### 1.1.1 HIGH-TEMPERATURE GAS/STEAM

#### 1.1.1.1 Gas Turbine (GT)

1. Introduction of gas turbine

In contrast to other combustion-type prime movers, the gas turbines emit much lower levels of NOx and $CO_2$ than other technologies for producing heating and electricity. It was in the 1950s that the micro gas turbine technology was first widely applied. Commercial microturbines only began to be deployed around 2000. The gas turbine operates based on the Brayton cycle. Figure 1.1 presents the schematic of a gas turbine cum CNG supply system. First, the air is compressed into the combustor and mixed with fuel. The mixed gas is combusted to drive the turbine to generate electricity. Then, the burned gas is used to increase the temperature of the compressed air and simultaneously produce hot water. The exhaust fuel gas is then dissipated into the ambient. From an energy conservation point of view, the dissipated superheated exhaust can be further exploited to produce heating or cooling. The temperature of compressed air is expressed as:

$$T_2 = T_1 + \Delta T = T_1 + \frac{T_1}{\eta_c} \left[ \left( \frac{P_2}{P_1} \right)^{\frac{\gamma_a - 1}{\gamma_a}} - 1 \right] \tag{1.1}$$

DOI: 10.1201/9781003343387-1

1

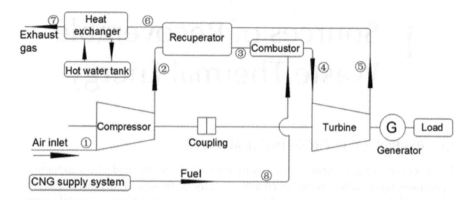

**FIGURE 1.1**  Schematic of a gas turbine system with CNG supply and heat recovery system [1]. Ambient air is compressed in the compressor. The compressed air flows into the combustor. Compressed natural gas is mixed with the compressed air in the combustor, resulting in a burning process. The burnt mixture is expanded into a turbine, generating electricity. Heat exchanges are used to harness thermal energy from exhausted gas flows for useful hot water production.

where $\eta_c$ is the isentropic efficiency.

The work consumed to drive the compressor is calculated as:

$$W_c = \dot{m}_1 c_{p,a} \Delta T_{12}$$

where $\dot{m}_1$ and $c_{p,a}$ are the mass flow rate and specific heat of air, respectively.

2. The energy balance equations of the combustor are expressed as:

$$\dot{m}_1 c_{p,a} T_3 + \dot{m}_f LHV = \dot{m}_4 c_{p,g} T_4 \tag{1.2}$$

where LHV is the low heat value of the fuel, and $c_{p,g}$ is the specific heat of superheated exhaust.

3. The temperature of exhaust is expressed as:

$$T_5 = T_4 - \Delta T_{45} = T_4 - \eta_t T_3 \left[ \left( \frac{1}{\pi_t} \right)^{\frac{\gamma_g - 1}{\gamma_g}} \right] \tag{1.3}$$

where $\gamma_g$ represents the heat capacity ratio of the superheated exhausted, and $\eta_t$ is the efficiency of the turbine.

The work provided by the turbine is expressed as:

$$W_t = \eta_m \dot{m}_4 c_{p,g} \Delta T_{45} \tag{1.4}$$

where $\eta_m$ is the mechanical efficiency.

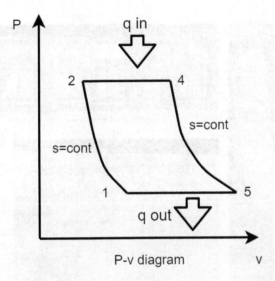

**FIGURE 1.2** Diagrams of an idealized Brayton cycle where $P$ is the pressure, $v$ is the volume, and $q$ is the heat added to or rejected by the system.

Figure 1.2 portrays the thermodynamic cycles of an idealized Brayton cycle. The Brayton cycle comprises four processes. During the isentropic process (1–2), the ambient air is compressed into the compressor. During the isobaric process (2–4), the fuel is mixed with compressed air and burned in the combustor. During the isentropic process (4–5), the heated, pressurized air is then expanded and passes through a turbine to generate electricity. During the isobaric process (5–1), the heat is rejected into the ambient.

4. Experiment case analysis of gas turbine
Figure 1.3 portrays a commercial C65-ICHP Capstone microturbine generating a net power of 65 kW. The characteristics of the gas turbine are listed in Table 1.1. The gas turbine comprises a compressor, a combustor, a turbine, and a recuperator. The micro gas turbine is fueled by CNG. To monitor its performance, mass flow meters along with RTD sensors are installed along the pipe systems to measure the flow rates and temperatures of the air, fuel, water, and exhaust gas. The ambient air at ① is first compressed to the pressure of the combustor (②) and then mixed with fuel fluxed from ⑧. The mixed gas is burned to drive the turbine at ④ to generate electricity. Thereafter, the burned gas is employed to increase the temperature of compressed air and produce hot water (⑥). The exhaust fuel gas, a byproduct from the turbine, is then dissipated into the ambient (⑦). In addition, gas leaking detectors are installed at both the CNG supply side and the experiment lab for safety reasons. The gas turbine is capable of operating under different power ratios. The performance of gas turbines under different load ratios is listed in Table 1.2.

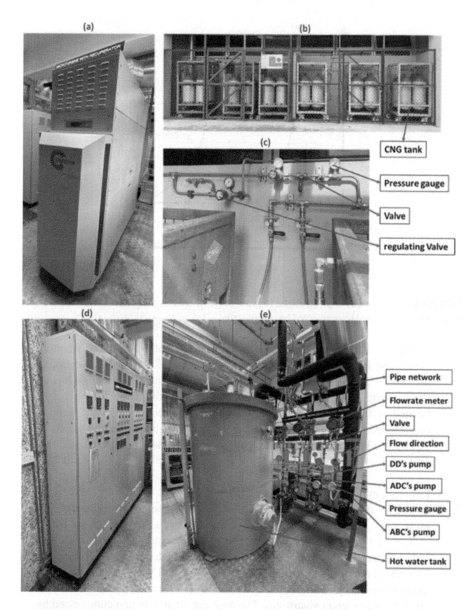

**FIGURE 1.3**  Pictorial view of the gas turbine and hot water storage system: (a) gas turbine; (b) CNG supply system; (c) pipe network, valves, and measuring gauges of the CNG system; (d) control system; and (e) thermally insulated hot water tank and its pumps, valves, gauges, and pipe network [1].

**TABLE 1.1**

**Specifications of the C65-ICHP Gas Turbine**

| Specifications | Value | Unit |
|---|---|---|
| Type | C65-ICHP Capstone microturbine | – |
| Electrical power output | 65 | kW |
| Voltage | 400/480 | VAC |
| Electrical service | 3-Phase, 4-wire wye | – |
| Frequency | 50/60 | Hz |
| Electrical efficiency LHV | 29% | – |
| Fuel inlet pressure | 75–80 | psig |
| Exhaust gas temperature | 309 | °C |
| Exhaust mass flow | 0.49 | kg/s |
| Hot water heat recovery | 122 | kW |

**TABLE 1.2**

**Key Parameters of the Gas Turbine Under Different Net Power Ratios**

| | Net Power Ratios (%) | | | | | | | |
|---|---|---|---|---|---|---|---|---|
| | 30 | 40 | 50 | 60 | 70 | 80 | 90 | 100 |
| $T_1$ (K) | 300.06 | 300.15 | 300.24 | 300.33 | 300.42 | 300.50 | 300.59 | 300.68 |
| $T_2$ (K) | 409.88 | 421.94 | 434.00 | 446.06 | 458.11 | 470.17 | 482.23 | 494.29 |
| $T_3$ (K) | 787.42 | 790.31 | 793.20 | 796.09 | 798.98 | 801.87 | 804.76 | 807.65 |
| $T_4$ (K) | 1,038.98 | 1,076.71 | 1,095.95 | 1,115.18 | 1,134.42 | 1,153.66 | 1,172.90 | 1,192.13 |
| $T_5$ (K) | 907.88 | 907.88 | 907.87 | 907.87 | 907.86 | 907.86 | 907.86 | 907.85 |
| $T_6$ (K) | 548.84 | 560.49 | 573.12 | 585.41 | 596.04 | 603.68 | 607.02 | 604.72 |
| $f(-)$ | 0.00760 | 0.00830 | 0.00884 | 0.00926 | 0.00956 | 0.00976 | 0.00988 | 0.00994 |
| $\dot{m}_g$ (kg/s) | 0.299 | 0.324 | 0.354 | 0.384 | 0.413 | 0.438 | 0.457 | 0.467 |
| $\dot{m}_a$ (kg/s) | 0.297 | 0.322 | 0.351 | 0.381 | 0.409 | 0.434 | 0.452 | 0.462 |
| $\dot{Q}_{rh}$ (kW) | 46.07 | 56.42 | 66.51 | 76.27 | 85.73 | 95.00 | 104.25 | 113.74 |
| $r_e(-)$ | 2.34 | 2.56 | 2.79 | 3.01 | 3.23 | 3.45 | 3.68 | 3.90 |
| $r_t(-)$ | 2.08 | 2.31 | 2.54 | 2.77 | 3.01 | 3.24 | 3.47 | 3.70 |
| $\varepsilon(-)$ | 0.77 | 0.77 | 0.77 | 0.78 | 0.78 | 0.78 | 0.78 | 0.78 |
| $\eta$ (%) | 20.44 | 22.13 | 23.72 | 25.00 | 26.18 | 26.95 | 27.62 | 28.00 |
| $W_t$ (kW) | 52.81 | 67.86 | 82.91 | 97.96 | 113.09 | 128.07 | 143.12 | 158.17 |
| $W_c$ (kW) | 32.28 | 40.49 | 48.70 | 56.91 | 65.12 | 73.33 | 81.54 | 89.75 |
| $P_e$ (kW) | 19.5 | 26 | 32.5 | 39 | 45.5 | 52 | 58.5 | 65 |

## 1.1.1.2 Internal Combustion Engine (ICE)

1. Introduction of internal combustion engine

   Internal combustion engines [2] are the most common type of heat engines. The first commercial internal combustion engine was invented by Étienne Lenoir around 1860 [3]. An ICE works by using either an Otto cycle or a diesel cycle. The comparisons of the two types of ICE are shown in Table 1.3. There are two types of reciprocating engines: spark ignition (SI) and compression ignition (CI). Diesel and heavy oil are the fuels used in compression ignition engines, which generate severe emissions. In contrast, spark ignition uses natural gas as the working fuel. An ICE generates electricity by expanding superheated gases with high pressure in its combustion chamber. During the exhaust gas purification process, waste heat is recovered via a heat exchanger. Figure 1.4 presents a schematic diagram of an internal combustion engine. In general, the fuel is burned to produce electricity. The heat exchangers produce hot water by harnessing thermal energy from exhausted gases. The waste heat from lube oil and water is recovered by the lube oil heat exchangers and jacket water heat exchangers. Accordingly, the engine is cooled.

   As illustrated in Figure 1.5a, an ideal diesel cycle comprises four processes. During the compression stroke (1–2), the air is adiabatically compressed in the cylinder. During the ignition isobaric heat addition, fuel is mixed with the compressed air under constant pressure (2–3). During the expansion stroke, the hot gases are adiabatically expanded in the cylinder (3–4). During the exhaust stroke, the hot gases are then dissipated (4–1).

   As illustrated in Figure 1.5b, an ideal Otto cycle consists of six processes. From point 1 to 0, the air is introduced into a cylinder at constant pressure. From point 1 to 2, the cylinder moves from the bottom to the top (adiabatic compression). From point 2 to 3, the volume is kept constant, and the heat is transferred from the external sources to the working fluid. From point 3 to 4, the hot gases are adiabatically expanded in the cylinder. From point 4 to 1, the heat is dissipated to the ambient. From 1 to 0, the air is dissipated into the ambient at constant pressure.

## TABLE 1.3
## Comparisons of Two Types of Internal Combustion Engines

| ICE | Petrol Engine | Diesel Engine |
| --- | --- | --- |
| Cycle type | Otto cycle | Diesel cycle |
| Ignition type | Air and fuel are mixed in a carburetor | Mixed in the cylinder |
| Compression ratio | Low | High |
| Power production | Low | High |
| Fuel consumption | High | low |
| Maintenance cost | Low | High |

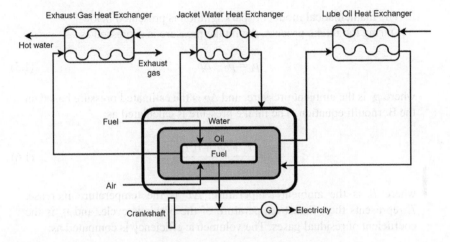

**FIGURE 1.4**  Schematic of an internal combustion engine.

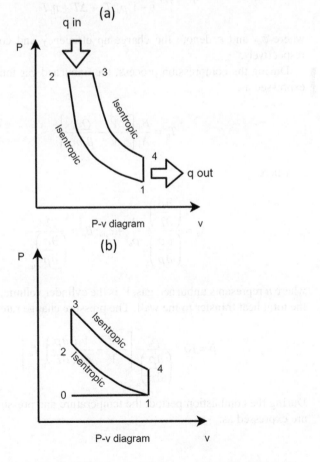

**FIGURE 1.5**  (a) Diagram of an ideal diesel cycle; (b) diagram of an ideal Otto cycle.

The mathematical model of spark ignition is presented below:
During the intake process, the intake pressure is expressed as:

$$p_i = p_0 - \Delta p \tag{1.5}$$

where $p_0$ is the ambient pressure, and $\Delta p$ is the estimated pressure based on the Bernoulli equation. The intake pressure is calculated as:

$$T_a = \frac{T_0 + \Delta T + \eta_r T_r}{1 + \eta_r} \tag{1.6}$$

where $T_0$ is the ambient temperature, $\Delta T$ is the temperature increase, $T_r$ represents the exhaust temperature of the previous cycle, and $\eta_r$ is the coefficient of residual gases. The volumetric efficiency is computed as:

$$\eta_v = \phi_{ed} \frac{\varepsilon}{\varepsilon - 1} \frac{p_a}{p_0} \frac{T_0}{T_0 + \Delta T + \eta_r T_r} \tag{1.7}$$

where $\phi_{ed}$ and $\varepsilon$ denote the charge-up efficiency and compression ratio, respectively.

During the compression process, the unburned gas temperature can be expressed as:

$$T_u = \frac{B_u}{A_u}\left[ -\frac{\dot{V}}{V} - \frac{\dot{Q}_{wu}}{B_u m_u} \right] \tag{1.8}$$

where

$$A_u = \frac{\left(\frac{\partial \rho}{\partial T}\right)_u}{\left(\frac{\partial \rho}{\partial p}\right)_u} \frac{1}{\rho_u} + c_{pu}; B_u = \frac{1}{\left(\frac{\partial \rho}{\partial p}\right)_u}$$

where $u$ represents unburned gas, $V$ is the cylinder volume, and $\dot{Q}_{wu}$ denotes the total heat transfer to the wall. The pressure change rate is expressed as:

$$\dot{p} = 10^{-5} \frac{\rho_u}{\left(\frac{\partial \rho}{\partial p}\right)_u}\left[ -\frac{\dot{V}}{V} - \frac{1}{\rho_u}\left(\frac{\partial \rho}{\partial T}\right)_u T_u \right] \tag{1.9}$$

During the combustion period, the temperature and pressure of burned gas are expressed as:

$$T_b = \frac{B_b}{A_b}\left[\frac{\dot{m}_b}{m_b}\left(1 - \frac{h_b}{B_b} - \frac{\dot{V}_b}{V_b} + \frac{1}{B_b m_b}(-Q_{wi} + m_b h_u)\right)\right] \qquad (1.10)$$

$$\dot{p} = 10^{-5}\frac{\rho_b}{\left(\frac{\partial\rho}{\partial p}\right)_b}\left[-\frac{\dot{V}_b}{V_b} - \frac{1}{\rho_b}\left(\frac{\partial\rho}{\partial T}\right)_b T_b + \frac{\dot{m}_b}{m_b}\right] \qquad (1.11)$$

where

$$A_u = c_{pb} + \frac{\left(\frac{\partial\rho}{\partial T}\right)_b}{\left(\frac{\partial\rho}{\partial p}\right)_b}\left[\frac{1}{\rho_u} - 10^{-5}\left(\frac{\partial h}{\partial p}\right)_b\right], B_u = \frac{1}{\left(\frac{\partial\rho}{\partial p}\right)_u}\left[\frac{1}{\rho_u} - 10^{-5}\left(\frac{\partial h}{\partial p}\right)_b\right].$$

During the expansion process, the temperature and pressure change rates are expressed as:

$$T_e = \frac{B_b}{A_b}\left[\frac{\dot{V}_b}{V_b} - \frac{Q_{wb}}{B_b m_b}\right] \qquad (1.12)$$

$$\dot{p} = 10^{-5}\frac{\rho_b}{\left(\frac{\partial\rho}{\partial p}\right)_b}\left[-\frac{\dot{V}_b}{V_b} - \frac{1}{\rho_b}\left(\frac{\partial\rho}{\partial T}\right)_b T_b\right] \qquad (1.13)$$

During the expansion process, the exhaust pressure and temperature are calculated as:

$$p_r = \left(\frac{1.05}{1.25}\right)p_0 \qquad (1.14)$$

$$T_r = \frac{T_b}{\left(\frac{p_b}{p_r}\right)^{\frac{1}{3}}} \qquad (1.15)$$

2. Experiment case analysis of internal combustion engine
   ICEs are widely deployed in various CCHP systems. Figure 1.6 depicts a pictorial view of the ICE with a power capacity of 130 kW [4]. The design parameters of the ICE are listed in Table 1.4. Natural gas is employed to drive the ICE for electricity and heating production. Research results have shown that the energy efficiency of the ICE is capable of reaching 60%. In addition, experimental research has found that increasing the air ratio can reduce the entropy generation of the ICE [4].

**FIGURE 1.6**  A pictorial view of ICE installed at Khajenasir University [4].

**TABLE 1.4**
**Design Conditions of the ICE [4]**

| Engine Specification | Value | Unit |
|---|:---:|:---:|
| Cylinder number | In-line 4, 4-stroke cycle | – |
| Cylinder bore | 128 | mm |
| Motor length | 2,650 | mm |
| Motor width | 1,000 | mm |
| Motor height | 1,500 | mm |
| Stroke | 166 | mm |
| Speed | 1,000 | rpm |
| Compression ratio | 12:1 | – |
| Maximum electrical power | 130 | kW |
| Weight | 1,850 | kg |
| Molar air-fuel ratio | 1.5 | – |
| Rated voltage | 220 | V |
| Rated current | 18 | A |
| Starting mode | 24 VDC electric starting system | – |
| Frequency | 50/60 | HZ |

### 1.1.1.3  Stirling Engine (SE)

1. Introduction of Stirling engine

   Stirling engine was invented by Robert Stirling in 1816 [5]. A Stirling engine [6] can cyclically compress and expand the working fluid at different temperatures. As portrayed in Figure 1.7, heat and cold sources are required to operate the Stirling engine. The heat produced from the fuel-burning process constitutes the heat source of the engine. A closed regenerative Stirling cycle continuously circulates the working fluid inside the engines. In other words, neither

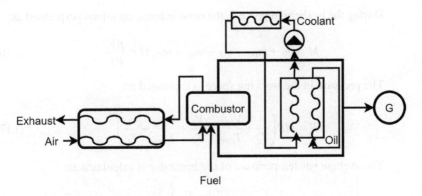

**FIGURE 1.7**  Schematic of a Stirling engine.

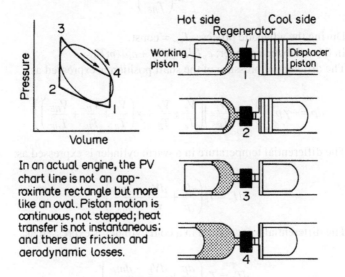

In an actual engine, the PV chart line is not an app-roximate rectangle but more like an oval. Piston motion is continuous, not stepped; heat transfer is not instantaneous; and there are friction and aerodynamic losses.

**FIGURE 1.8**  Four working processes of SE [5].

the working fluid nor any internal parts are contacted by the combustion products. Stirling engines can use landfill gas as their primary energy source [6]. Figure 1.7 provides a schematic perspective of a Stirling engine. The Stirling engine comprises a combustor, a generator, and two heat exchangers. One of the heat exchangers uses the heat from the exhausted gas to heat the water. The other employs a coolant to cool down the Stirling engine.

The ideal working processes of the Stirling engine include isothermal compression, isochoric heating, isothermal expansion, and isochoric cooling. The four working processes are illustrated in Figure 1.8 [5]. During the isothermal expansion, the working gas is heated by external sources. During the constant-volume heat-removal process, the gas is cooled down in the generator. During the isothermal compression, the heat is transferred from the working gas to the cold sink. During the constant-volume heat-addition process, the working gas is heated in the generator.

During the isothermal process, the mass balance equation is expressed as:

$$M = m_c + m_{HC} + m_R + m_{HE} + m_E, M = \frac{pV}{RT}$$ (1.16)

The pressure in the working space is expressed as:

$$P = MR\left(\frac{V_c}{T_c} + \frac{V_{HXC}}{T_{HXC}} + \frac{V_R}{T_R} + \frac{V_{HXE}}{T_{HXE}} + \frac{V_E}{T_E}\right)^{-1}$$ (1.17)

The average gas temperature of the generator is calculated as:

$$T_R = \frac{(T_{HC} - T_{HE})}{\ln\left(\dfrac{T_{HC}}{T_{HE}}\right)}$$ (1.18)

During the adiabatic process, $T_{gas}$ = const.
In the adiabatic cylinder: $c_p T d_m = dW + c_v d(mT)$.
The differential pressure of the shaft position is expressed as:

$$dp = -\gamma p\left(\frac{dV_c}{T_c} + \frac{dV_E}{T_E}\right)\left(\frac{V_c}{T_c} + \frac{V_E}{T_E} + \gamma\left(\frac{V_R}{T_R} + \frac{V_{Hc}}{T_{Hc}} + \frac{V_{HE}}{T_{HE}}\right)\right)^{-1}$$ (1.19)

The differential temperature in a warm cylinder is expressed as:

$$dT_c = T_c\left(\frac{dp}{p} + \frac{dV_c}{V_c} - \frac{dm_c}{m_c}\right)$$ (1.20)

The differential temperature in a cold cylinder is expressed as:

$$dT_E = T_E\left(\frac{dp}{p} + \frac{dV_E}{V_E} - \frac{dm_E}{m_E}\right)$$ (1.21)

2. Experiment case analysis of Stirling engine
   Stirling engines are promising prime movers for residential Combined
   Cooling, Heat, and Power (CCHP) applications due to their low emission,
   maintenance, noise, vibration, and high thermal-to-electrical efficiency
   [6]. Thorsen et al. [7] tested a 3-kW Stirling engine to supply electricity
   and heating for a single-family house in 1996. In 2001, a 10-kW Stirling
   engine was developed and tested. The electrical efficiency was reported to
   be 24% [6]. Figure 1.9 presents a pictorial view of a Stirling engine with a
   capacity of 3.9 kW [8]. Key research results have revealed that the working
   fluid's pressure loss at the generator is more than twice that at the cooler
   (Table 1.5).

**FIGURE 1.9** A pictorial view of SE [8].

**TABLE 1.5**
**Design Conditions of the SE [8]**

| Engine Specification | Value | Unit |
|---|---|---|
| Working fluid | Helium | – |
| Designed speed | 1,500 | RPM |
| Mean pressure | 2 | MPa |
| Indicated power output | 3.9 | kW |
| Thermal efficiency | 0.26 | – |
| Displacer/piston swept area | 81.71 | cm$^2$ |
| Displacer stroke | 5.65 | cm |
| Piston stroke | 4.44 | cm |

### 1.1.1.4 Steam Turbine (ST)

High-pressure steam is employed to operate a steam turbine [9] to generate electricity. Charles Parsons invented the modern steam turbine in 1884 [10]. Steam turbines are large-scale applications based on the Rankine cycle. A boiler, a steam turbine, a condenser, and a pump constitute the key components of a steam turbine. A boiler generates steam, which is then expanded into a turbine. The steam turbine can be used to recover waste steam energy and produce electricity. Figure 1.10 presents the schematic of two typical steam turbines.

As illustrated in Figure 1.11, a cycle includes four processes. From point 1 to 2, the working fluid is first pumped to high pressure. From point 2 to 3, the high-pressure working fluid is heated at constant pressure, and saturated vapor is generated in the boiler. From point 3 to 4, the generated vapor is expanded through a turbine for

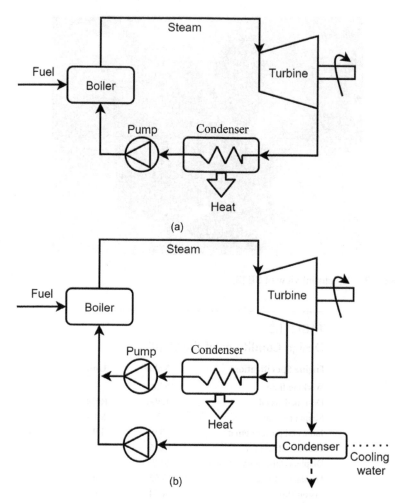

**FIGURE 1.10** Schematic diagram of (a) a back-pressure steam turbine and (b) an extraction-condensing steam turbine.

electricity generation. From 4 to 1, the wet vapor is condensed at constant pressure. For an ideal Rankine cycle, processes 1–2 and 3–4 are isentropic processes.

The mechanical power is expressed as:

$$\frac{dW}{dt} = V\frac{d\rho}{dt} = F_{in}(t) - F_{out}(t) \tag{1.22}$$

where $W$ is the weight of steam, and $F$ represents the mass flow rate of steam. The steam flow is assumed to be proportional to the steam pressure in the turbine:

$$F_{out} = P\frac{F_o}{P_o} \tag{1.23}$$

where $F_o$ and $P_o$ represent the rated flow rate and pressure in the turbine, respectively.

Steam turbines have been widely deployed globally as prime movers. Coal, natural gas, and geothermal are the primary energy sources for steam turbines [11].

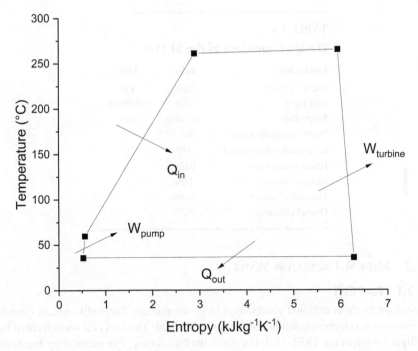

**FIGURE 1.11**  Diagram of an ideal Rankine cycle. The operating pressures range from 0.06 to 50 bar.

**FIGURE 1.12**  A pictorial view of a steam turbine [13].

Compared to internal engines, steam turbines are capable of providing power spanning 100 kW to 250 MW [12]. Reports have documented that there are 699 sites in the United States with installed steam turbines in cogeneration systems [13]. Figure 1.12 presents a pictorial perspective of a steam turbine. The design parameters of the steam turbine are listed in Table 1.6. High total efficiency and low power to heat are two features of steam turbines in a cogeneration system.

**TABLE 1.6**
**Design Conditions of the ST [13]**

| Parameters | Value | Unit |
|---|---|---|
| Electric power | 3,000 | kW |
| Fuel input | 208 | MMBtu/h |
| Steam flow | 152,600 | lbs/h |
| Steam inlet temperature | 301.7 | °C |
| Steam outlet temperature | 189 | °C |
| Power to heat ratio | 0.066 | – |
| Electric efficiency | 4.9% | – |
| Thermal efficiency | 74.8% | – |
| Overall efficiency | 79.7% | – |

## 1.1.2 MEDIUM-TEMPERATURE WATER

### 1.1.2.1 Fuel Cell

In contrast to the traditional combustion in prime movers, fuel cells convert chemical energy into electricity while producing waste heat. The fuel cell was invented by William Grove in the 1830s [14]. For more than a century, this technology has been relatively underappreciated. Recently, FC technologies have been at the forefront of research due to their capabilities to achieve high efficiency while producing low emissions. In contrast to batteries, which can only store a limited amount of energy, the FC technology can produce both electricity and heat simultaneously. Figure 1.13 depicts the schematic diagram of a fuel cell. Basically, fuel is fed to the anode, while air is fed to the cathode. At the anode side, hydrogen molecules are separated into protons and electrons by a catalyst. The electrons are then passed through an external electrical circuit. As a result, electricity is generated. To further improve electricity generation efficiency, fuel cells can incorporate an Organic Rankine Cycle (ORC) as the bottoming coupling technology to yield a hybrid system [15].

The reactions of a proton exchange membrane fuel cell are briefly introduced as follows:

$$\text{Anode}: H_2 \rightarrow 2H^+ + 2e^- \tag{1.24}$$

$$\text{Cathode}: 2H^+ + 2e^- + 0.5O_2 \rightarrow H_2O \tag{1.25}$$

The overall reaction can be expressed as:

$$H_2 + 0.5O_2 \rightarrow H_2O \tag{1.26}$$

The open circuit voltage is estimated as [16]:

$$E = 1.229 - 0.8*10^{-3}(T_{fc} - 298.15) + 4.3085*10^{-5}T_{fc}\ln(P_{H_2}P_{O2}^{0.5}) \tag{1.27}$$

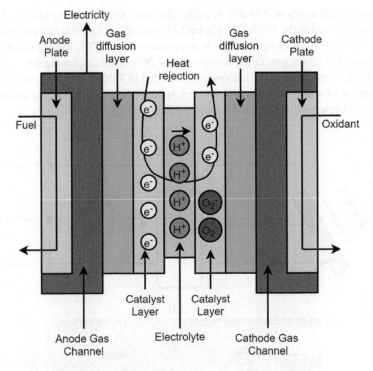

**FIGURE 1.13** Schematic diagram of a fuel cell.

where $P_{H2}$ and $P_{O2}$ are the pressure of hydrogen and oxygen, respectively.

$$P_{H_2} = 0.5 P_{H_2O}^{sat} \left( \frac{1}{\exp\left(\dfrac{1.653i}{T_{fc}^{1.334}}\right) x_{H_2O}^{sat}} - 1 \right) \tag{1.28}$$

$$P_{O_2} = P\left(1 - x_{H_2O}^{sat} - x_{N_2}^{channel} \exp\left(\frac{0.291i}{T_{fc}^{0.832}}\right)\right) \tag{1.29}$$

where $i$ and $P$ are the current density and pressure, respectively. $P_{H_2O}^{sat}$ is the saturation pressure and is expressed as follows, according to Miansari Me et al. [17]:

$$\log P_{H_2O}^{sat} = -2.1794 + 0.02953(T_{fc} - 273.15) - 9.1837 * 10^{-5}(T_{fc} - 273.15)^2$$
$$+ 1.4454 * 10^{-7}(T_{fc} - 273.15)^3 \tag{1.30}$$

### 1.1.2.2 Photovoltaic/Thermal (PVT) Hybrid Solar System

Photovoltaic/Thermal (PVT) hybrid solar system is a promising technology that utilizes solar energy to simultaneously generate electricity and useful heating for buildings and industrial applications. In contrast to conventional PV modules, PVT is able to achieve better energy efficiency as it is able to utilize the solar spectrum more effectively. The PVT was proposed in the 1970s [18]. Figure 1.14 presents a schematic and pictorial view of the PVT system. In comparison with amorphous silicon PVT, experimental results have shown that power generation is improved by 20% when employing the crystalline silicon PVT. The system specifications are tabulated in Table 1.7. Due to the efficiency improvement of solar collectors and the corresponding cost reduction, the deployment of PVT has been gaining traction for many commercial applications.

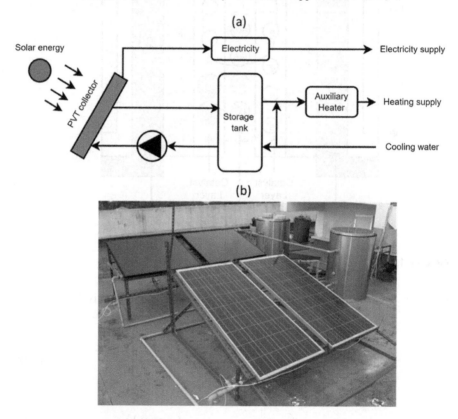

**FIGURE 1.14**    (a) Diagram of a PVT system and (b) a pictorial view of a PVT system [19].

**TABLE 1.7**
**Specification of the Experiment PVT System**

|  | Amorphous Silicon PVT | Crystalline Silicon PVT |
| --- | --- | --- |
| Collector area (m²) | 0.9 | 1.96 |
| Packing factor of PVT collector | 0.9 | 0.9 |
| Working fluid thermal capacitance (kJ/kg/K) | 4.19 | 4.19 |
| Nominal efficiency of PV panel (%) | 6.3 | 12.7 |
| Collectors slope (degree) | 15 | 15 |
| Glass cover transmittance | 0.92 | 0.92 |
| Storage tank volume (m³) | 0.31 | 0.31 |

### 1.1.3 SUMMARY OF THE PRIME MOVER

There are many applications for prime movers in both commercial and residential sectors. Prime movers generate recoverable thermal waste by means of exhaust, including superheated steam/gas, hot water, and solid carriers. Depending on the types and temperatures of waste heat, a heat recovery system can be incorporated to produce cooling, heating, or electricity. The typical types of prime movers and their features are listed in Table 1.8.

**TABLE 1.8**
**Typical Types of Different Prime Movers**

|  | ICE [20] | SE [21] | GT [22] | ST [12] | FC [23] |
|---|---|---|---|---|---|
| Capacity (kW) | 10–18,000 | 1–55 | 500–300,000 | 100–250,000 | 5–1,400 |
| Cycle | Otto/diesel cycle | Stirling cycle | Brayton cycle | Rankine cycle | – |
| Electric efficiency (%) | 27–41 | 20–35 | 22–36 | 6.27–7.31 | 35–42.5 |
| Fuel type | NG, biogas, propane | All | NG, biogas, propane, oil | All | Hydrogen, NG, propane, methanol |
| Thermal product | Hot water, low-pressure steam | – | Hot water, low-pressure steam, high-pressure steam | Low-pressure steam, high-pressure steam | Hot water, low-pressure steam, high-pressure steam |
| Temperature of dissipated heat (°C) | 350–648 | – | 380–550 | 147–192 | 371 |
| Exhaust flow (kg/h) | 0.5–55 | – | 67–474 | 9,094–224,285 | – |
| Advantages | Fast start-up | Low noise and maintenance | High-quality thermal output | Lifetime is very long | Low emission |

## 1.2 BYPRODUCTS FROM INDUSTRIAL ACTIVITIES

### 1.2.1 HIGH-TEMPERATURE SOLID GRANULAR

High-temperature solid granulars (HTSG) are massively produced all over the world in chemistry, metallurgy, and building materials industries [24]. These HTSG include cement with a temperature as high as 1,573 K, coke and slags with a temperature close to 1,323 K, and iron-ore sinters reaching a temperature of 1,823 K. The annual productions of HTSG in China, the United States, and India are 4.29 billon tons,

0.24 billon tons, and 0.49 billion tons, respectively [24]. Thus, a significant amount of fuel consumption and $CO_2$ emission can be reduced if heat recovery technologies are designed and deployed to harness the thermal energy from HTSG effectively.

### 1.2.1.1  Cement Clinker/Grate Clinker Cooler

Construction materials are widely used all over the world. Cement is used to build concrete, which is a mixture of cement, aggregates, and water. Globally, concrete is the most widely used manufacturing product, with over three tons produced per person per year. In most cases, concrete can replace other building materials such as wood, steel, plastic, aluminum, and wood because it is more effective, less expensive, and more efficient than all other materials combined. The popularity of concrete as a building material stems from its low cost of manufacture, its ability to be manufactured locally from widely available raw materials, its moldability, and its high compressive strength.

Figure 1.15 illustrates the cement clinker grate cooler for waste heat recovery processes. In a cement plant, grate coolers are employed to provide heated air for the rotary kiln and preheat air for the rotary kiln during the clinker manufacturing process. There are three types of grate coolers: rotary grate coolers, traveling grate coolers, and reciprocating grate coolers. Today, most cement plants use reciprocating grate coolers to cool clinker, while the first two are already obsolete.

It is possible to convert useful thermal energy from the kiln exhaust streams into electricity via the kiln cooler and kiln preheater system. Heat is emitted directly into the atmosphere by clinker coolers at temperatures spanning 250–340°C. On the kiln's charge side, the kiln gas is employed to dry materials in raw mills and/or coal mills at a temperature ranging from 300 to 400°C before it is filtered through electrostatic precipitators or bag filter houses for final disposal. During downtimes, dust collectors would cool the exhaust gases with water sprays or cold air. The majority of waste heat recovery projects have been used for power generation at cement plants because cement plants have no significant low-temperature heating requirements. A kiln's design, production, and moisture content affect the amount of waste heat available for recovery and the amount of heat required for drying in the raw mill, solid fuel system, and cement mill. It is estimated that almost 30% of electricity consumption can be reduced when the waste heat can be effectively harnessed.

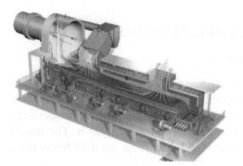

**FIGURE 1.15**  Cement clinker grate cooler for waste heat recovery [25].

For cement kilns, the waste heat recovery power system operates based on the Rankine Cycle, which is the basis for conventional thermal power generating stations. As described in the previous section, Rankine cycles feature a heat source (boiler) that converts liquids into steam, which is then expanded in a generator to generate electricity. A boiler feedwater pump returns the condensate from the condenser so that the low-pressure vapor exhaust from the turbogenerator can be cooled back to its liquid state.

Steam Rankine cycle, Organic Rankine cycle, and Kalina cycle are the three primary waste heat recovery technologies commonly applied in the cement industry. Steam Rankine cycles are most commonly used in the cement industry and are typically preferred when a source heat temperature exceeds 300°C. There are a few cement kilns that have been equipped with ORC systems. In the cement industry, the incorporation of the Kalina cycle is now being piloted.

Power systems for reusing waste heat from the cement industry have been introduced by several Japanese companies. For instance, Sumitomo Osaka Cement incorporated the first waste heat recovery system installed by Kawasaki Heavy Industries (KHI) in 1980 [26]. At Taiheiyo Cement's Kumagaya plant, the first 15 MW commercial system has been operating since 1982 [27]. By 2012, over 700 units are operating in China as a result of government policies and Clean Development Mechanism (CDM) incentives. The market is chiefly dominated by Asian companies; most of the suppliers are Chinese. The latest generation of waste heat recovery systems offers up to 45 kWh/t of clinker heat recovery thanks to supercritical steam parameters and improved efficiencies.

As illustrated in Figure 1.16, the waste heat boilers installed in waste heat recovery systems on New Suspension Process (NSP) kilns are employed to create

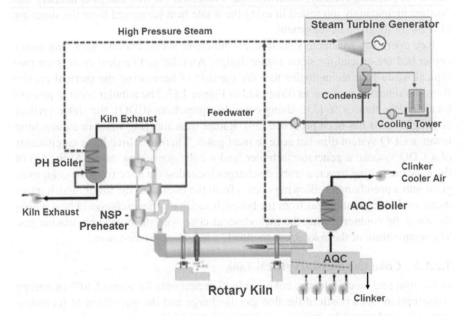

**FIGURE 1.16**   Waste heat recovery system on an NSP cement kiln [28].

medium/low-pressure steam from hot exhaust streams that are exiting the preheaters and air quench clinker cooler (AQC). An electric generator is powered by a condensing steam turbine that feeds the steam into a condensing steam turbine. Boilers are used to recycle waste heat and receive hot condensate from the condenser. Steam turbine generators, condensers, water treatment systems, boiler feed pumps, recooling systems, and ancillary equipment make up the entire system. An NSP kiln system typically produces an exhaust with temperatures ranging from 280 to 450°C, depending on how many preheater stages are included [28]. Depending on the cooling air volume and recuperation efficiency, the waste air from the clinker cooler has temperatures ranging from 250 to 330°C [28].

### 1.2.1.2   Sinter/Annular and Vertical Cooler

The process of producing steel produces a large number of waste heat resources globally. With 21.91% of the total waste heat generated, the residual heat generated during the sintering process is second only to the waste heat generated by the iron-making process [29]. Two processes are involved in sintering: sintering machine sintering and cooling machine cooling. As much as 13–23% of the sintering process energy consumption is consumed by the residual heat taken away by the sintering part of the flue gas. Sintering consumes approximately 20–25% of the energy from the residual heat taken away by the cooling machine flue gas [30]. It is apparent that the cooling process of the cooling machine generates a large amount of waste heat which should be recycled.

The iron and steel industry has applied a number of advanced technologies to reduce energy consumption [31], but the amount of energy consumed still is prominent in the entire industry, especially in the sintering process [32]. Further, waste heat from sintering remains underutilized. Therefore, the iron and steel industry has become increasingly interested in using the waste heat harnessed from the sintering process for its green development.

Key operating parameters such as air flowrate, actual cooling area, and sinter cooler bed are crucial for sinter cooler design. Annular and vertical coolers are two typical waste heat technologies that are capable of harnessing the thermal energy from the sintering process, as illustrated in Figure 1.17. The annular cooling process takes up less than 30% [33]. Using coke dry quenching (CDQ), the sinter vertical cooling process has been proposed [34]. Rather than spraying water in a quenching tower, a CDQ system dips hot coke in inert gases. There are three basic components of a CDQ system: a generator, a boiler, and a coke quenching tower. At the top of the coke quenching tower, a crane discharges incandescent coke from the coke-oven plant into a pre-chamber. Blowing occurs from the bottom of the tower, which introduces cooled circulation gas from the boiler. It facilitates the exchange of heat within the tower by counterflowing the incandescent coke with the cooled circulation gas. The temperature of the coke drops to 200°C when it leaves the tower.

### 1.2.1.3   Coke Oven Coke/Vertical Tank

In the iron and steel industry, coke production accounts for around 10% of energy consumption. As a result of the flue gas discharge and the quenching of incandescent coke and raw coke oven gas, a large amount of thermal energy is lost during

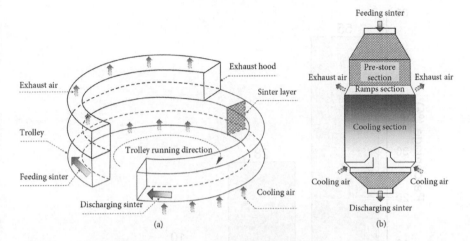

**FIGURE 1.17** Diagram of (a) annular cooling process and (b) vertical cooling process [35].

the coke production process [36]. Approximately 37–48% of the thermal energy is lost from the incandescent coke at 1050°C, while 24–36% is lost from the raw coke oven gas at 800°C. In order to reduce $CO_2$ emissions, coking plants should employ heat recovery systems [37]. The waste heat system designed for the coke plant consists of two subsystems. The first one is a coke dry quenching integrated with a steam cycle. During this process, recirculating of the inert gas is carried out by passing it upward through the quenching chamber, and it cools the hot coke from 1050°C to 200°C. A series of heat exchangers serve as a steam generator by working with an inert gas that enters the bottom of the quenching chamber at 175°C and exits at 945°C. After the surface condenser exhausts its condensed water, it is pumped to a deaerator that heats the water to 104°C. In the economizer, 230°C is achieved by heating the boiler feed water from the deaerator. In order to generate superheated intermediate-pressure steam, the make-up boiler feed water goes to an intermediate-pressure steam generator and two intermediate-pressure steam superheaters [36].

#### 1.2.1.4 Slag from Steel Industry

Steel is one of the most energy-intensive industries, accounting for approximately 4–5% of global energy consumption. Products, molten slag, and waste gases are the main sources of high-temperature waste heat available from the iron and steel industry. A byproduct of steelmaking is molten slag, which is emitted at extremely high temperatures. The temperature of molten slag is about 1,450–1,550°C. Molten slag includes blast furnace slag and steel slag. As illustrated in Figure 1.18, molten slag accounts for 35% of waste heat [38]. As a result of their low thermal conductivity, inside crystallization, and discontinuous availability, the process of heat recovery from slags presents several fundamental challenges. The traditional heat recovery technology of water quenching involves using cold water to cool down slag [39]. Some chemical methods have also been studied for their potential to harness waste heat for commercial applications [40].

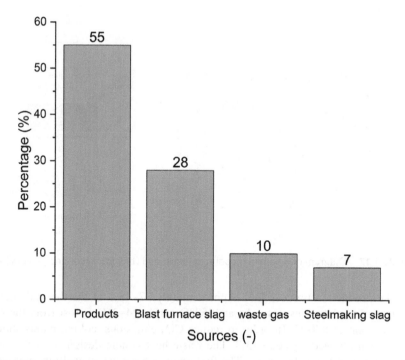

**FIGURE 1.18**    High-temperature waste heat sources from the steel industry [38].

## 1.2.2   HIGH-TEMPERATURE GAS/STEAM/WATER

### 1.2.2.1   Exhaust from Aluminum Industry

Aluminum is the second most produced non-ferrous metal after steel, having a higher volume of production than all other non-ferrous metals combined [41]. Figure 1.19 briefly portrays the thermal energy loss during the aluminum smelting process. Bauxite mining, alumina refining, and aluminum smelting are the three major stages in the production of aluminum. Waste heat accounts for half of the energy lost during aluminum production. The aluminum smelter dissipates about 30–45% of its total waste heat [42]. Organic Rankine cycle, heat supply system, and combined heat and power system are typical methods employed to harness the waste heat from the aluminum industry [42].

### 1.2.2.2   Exhaust from Nuclear Power Plant

Research results have highlighted that around 2/3 of the generated heat is dissipated into the ambient during nuclear power generation processes [44]. International Energy Agency estimates that the total electricity generated by nuclear power plants will reach 17,844 GWe by 2050 [45]. Therefore, it is imperative to harness the available waste heat from nuclear power plants for energy saving. Figure 1.20 illustrates the schematic flow of a nuclear reactor. The generated steam can be used for district heating or power generation [46]. Typically, a thermodynamic Rankine cycle is installed to harness the thermal energy from the nuclear reactor.

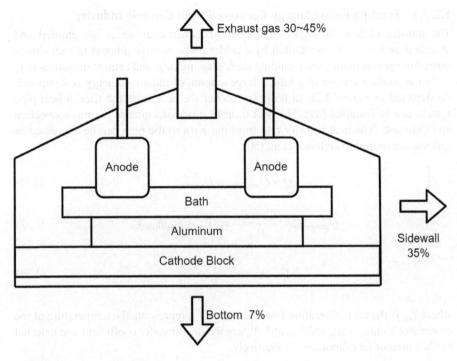

**FIGURE 1.19** Thermal energy loss during the aluminum smelting process [43].

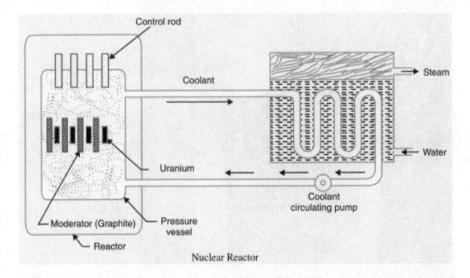

**FIGURE 1.20** Diagram of a nuclear reactor [47]. A huge amount of heat is released when the reactor works. Coolant solution must be used to cool down the reactor. The superheated coolant is then employed to heat up water to generate steam.

### 1.2.2.3  Heat Radiation Energy Recovered from Ceramic Industry

The transfer of heat through radiation occurs when heat waves are emitted and absorbed, reflected, or transmitted by a colder body. A large amount of radiation is wasted in various industries, including steel, glassmaking, and cement industries [48].

In the cooling section of a kiln, a large amount of radiative energy is dissipated. As depicted in Figure 1.21, in order to recover the heat from the tiles, a heat pipe system can be installed [49]. The heat transfer processes include natural convection and radiation. The heat transfer rate from the Kiln to the pipe can be expressed as follows, according to Delpech et al. [50]:

$$\dot{Q} = \dot{Q}_{convevtion} + \dot{Q}_{radiation} \tag{1.31}$$

$$\dot{Q}_{convevtion} = \frac{T_{air} - T_{ow}}{R_{con}} = (T_{air} - T_{ow})h_{con}A_{es} \tag{1.32}$$

$$\dot{Q}_{radiation} = \frac{T_{heater} - T_{ow}}{R_{rad}} \tag{1.33}$$

where $T_{air}$ is the air temperature inside the kiln, $T_{ow}$ represents the temperature of the evaporator's outer wall, and $h_{con}$ and $A_{es}$ are the heat transfer coefficient and external surface area of an evaporator, respectively.

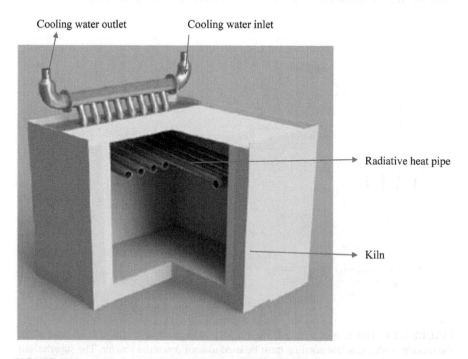

**FIGURE 1.21**   Radiative heat pipes in a ceramics kiln [49].

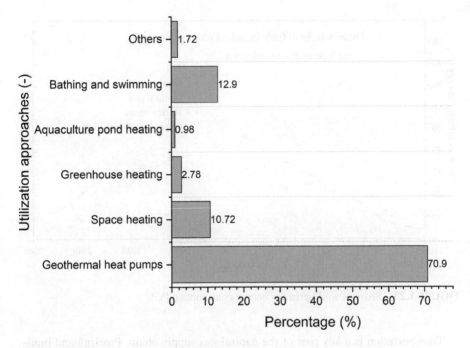

**FIGURE 1.22** Direct use of geothermal application worldwide in 2015 [51].

### 1.2.2.4 Geothermal Energy Utilization

Direct use of geothermal energy is one of the most versatile forms of utilizing its energy. Direct use of geothermal energy includes geothermal heat pumps, space heating, greenhouse heating, aquaculture pond heating, etc. Figure 1.22 shows the distributions of various geothermal energy utilization. The annual energy use of geothermal energy by heat pumps is 326,848 TJ/year [51]. Geothermal energy is utilized to produce steam. A Rankine cycle is installed to harness the steam to generate electricity. Power generation from convective hydrothermal systems is one of the most common geothermal resources being exploited worldwide [52].

### 1.2.3 COLD ENERGY FROM LNG REGASIFICATION PROCESS

Coal, oil, and natural gas are the most well-known fossil fuels. For much of the past 100 years or so, the world's primary energy mix has been dominated by a single energy source. Before the 1960s, coal is the most popular fossil fuel. Up to now, oil has overtaken that prime position. However, natural gas is the only fossil fuel with a rising share in the past ten years and is expected to be the most consumed form of fossil fuel by 2050 under the accelerated scenario [53], as highlighted in Figure 1.23. Natural gas causes the least amount of emissions, such as sulfur dioxide, carbon dioxide, and nitrogen oxide, per unit of heat release when it is compared to oil or coal. Further, natural gas also presents other appealing advantages, such as efficient combustion, reliable and durable supply, flexible implementations in vehicles, power plants, and industrial and residential sectors.

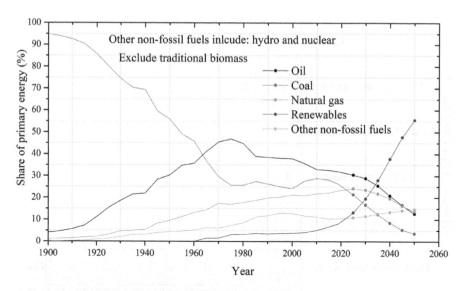

**FIGURE 1.23**   Share of world primary energy consumption [53].

Transportation is a key part of the natural gas supply chain. Pipeline and liquified natural gas (LNG) are the two commercial, mature, and profitable methods to transport natural gas. The pipeline is the oldest method to deliver natural gas but is still deemed to be the most suitable technology for distances less than 1,000 km. In contrast, the transportation of natural gas in the form of LNG will be more profitable for distances larger than 3,500 km [54]. When the natural gas is cooled down to −162°C under atmospheric pressure, it becomes a cryogenic liquid, known as LNG. The density of LNG ranges from 400 to 500 kg/m³. In other words, the volume is reduced to around 1/620 when the same quality of natural gas is transported and stored. Therefore, LNG has become an economical and feasible medium for transportation and storage, even if it requires a relatively large amount of energy to liquefy natural gas.

Production, liquefication, transportation, and regasification are the four key steps in the LNG supply chain. Regasification is the latest step before natural gas is sent to consumers. LNG is first stored in the storage tank at the receiving terminal and then sent to the vaporizers to produce natural gas. The common vaporizers include open rack vaporizers (ORV), submerged combustion vaporizers (SCV), shell and tube vaporizers (STV), intermediate fluid vaporizers (IFV), and ambient air vaporizers (AAV). The ORV and SCV are the most common vaporizers and occupy around 70% and 20% shares, respectively [55]. The schematic diagram of the LNG regasification process in these two vaporizers is shown in Figure 1.24. During the regasification process, LNG releases a large amount of cold energy (860 kJ/kg). Through the vaporizers, the high-quality LNG cold energy, if not harvested, will be wasted. For instance, the LNG cold energy is dumped into the ocean through the seawater and becomes an environmental hazard when using the ORV to regasify the LNG. Even extra fuel is also required to heat the water in the SCV.

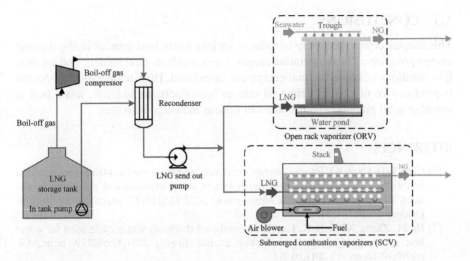

**FIGURE 1.24** Schematic diagram of the LNG regasification process.

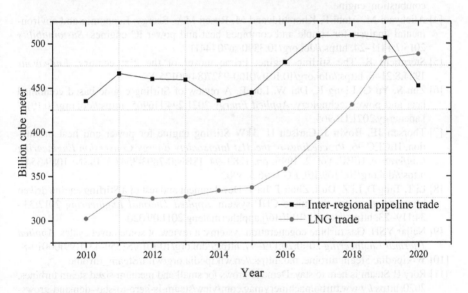

**FIGURE 1.25** Variation of LNG trade and pipeline trade in recent years [56].

In recent years, the LNG trade has projected a more robust growth than the pipeline trade due to market flexibility and energy independence issues, as presented in Figure 1.25 [56]. In 2010, the total natural gas trade was about 737.7 billion cube meters, and only 41% (338.8 billion cube meters) of the natural gas was traded in the form of LNG. However, the share of LNG trade exceeded the pipeline trade and occupied about 52% in 2020. According to the current global LNG trade quantity, the available cold energy reaches as high as 9211 MW. Hence, the recovery of cold energy during the LNG regasification process is an intriguing subject that has huge potential to recover waste thermal energy and mitigate carbon emissions.

## 1.3 CONCLUSIONS

This chapter comprehensively introduces various waste heat sources in the thermal energy processes. The temperature ranges, types, mathematical models, and application situations of waste thermal energy are introduced. High-temperature gas/steam is produced from various prime movers as byproducts. In addition, waste heat is stored in solid granular and hot water in various industrial activities.

## REFERENCES

[1] Chen WD, Chua KJ. Energy, exergy, economic, and environment (4E) assessment of a temperature cascading multigeneration system under experimental off-design conditions. *Energy Conversion and Management* 2022;253:115177. https://doi.org/10.1016/j.enconman.2021.115177

[2] He M, Zhang X, Zeng K, Gao K. A combined thermodynamic cycle used for waste heat recovery of internal combustion engine. *Energy* 2011;36:6821–9. https://doi.org/10.1016/j.energy.2011.10.014

[3] Wikipedia. Internal combustion engine. n.d. https://en.wikipedia.org/wiki/Internal_combustion_engine.

[4] Aliehyaei M, Atabi F, Khorshidvand M, Rosen MA. Exergy, Economic and environmental analysis for simple and combined heat and power IC engines. *Sustainability* 2015;7:4411–24. https://doi.org/10.3390/su7044411

[5] Sternlicht B. The stirling engine: Prime mover of the 21st century. *Endeavour* 1984;8:21–8. https://doi.org/10.1016/0160-9327(84)90125-X

[6] Zhu S, Yu G, Liang K, Dai W, Luo E. A review of Stirling-engine-based combined heat and power technology. *Applied Energy* 2021;294:116965. https://doi.org/10.1016/j.apenergy.2021.116965

[7] Thorsen JE, Bovin J, Carlsen H. 3 kW Stirling engine for power and heat production. IECEC 96. *Proceedings of the 31st Intersociety Energy Conversion Engineering Conference*, IEEE, vol. 2, 1996, pp. 1289–94. ISBN:0-7803-3547-3, ISSN: 1089-3547. https://doi.org/10.1109/IECEC.1996.553902

[8] Li T, Tang D, Li Z, Du J, Zhou T, Jia Y. Development and test of a Stirling engine driven by waste gases for the micro-CHP system. *Applied Thermal Engineering* 2012;33–34:119–23. https://doi.org/10.1016/j.applthermaleng.2011.09.020

[9] Najjar YSH. Gas turbine cogeneration systems: a review of some novel cycles. *Applied Thermal Engineering* 2000;20:179–97. https://doi.org/10.1016/S1359-4311(99)00019-8

[10] Wikipedia. Steam turbine, n.d. https://en.wikipedia.org/wiki/Steam_turbine

[11] Rory P. Steam is here to stay. Demand grows for small and medium-sized steam turbines, 2020. https://www.turbomachinerymag.com/view/steam-is-here-to-stay-demand-grows-for-small-and-medium-sized-steam-turbines

[12] U.S. Environmental Protection Agency, Combined Heat and Power Partnership. *Technology Characterization: Steam Turbines*. Washington, D.C.: United States Environmental Protection Agency; 2015.

[13] USA Department of Energy. Steam turbine; n.d. chrome-extension://efaidnbmnnnibpcajpcglclefindmkaj/https://www.energy.gov/sites/default/files/2016/09/f33/CHP-Steam%20Turbine.pdf

[14] Appleby AJ. From Sir William Grove to today: Fuel cells and the future. *Journal of Power Sources* 1990;29:3–11. https://doi.org/10.1016/0378-7753(90)80002-U

[15] Liu L, Yang Q, Cui G. Supercritical carbon dioxide(s-$CO_2$) power cycle for waste heat recovery: A review from thermodynamic perspective. *Processes* 2020;8, 1461. https://doi.org/10.3390/pr8111461

[16] Mann RF, Amphlett JC, Hooper MAI, Jensen HM, Peppley BA, Roberge PR. Development and application of a generalised steady-state electrochemical model for a PEM fuel cell. *Journal of Power Sources* 2000;86:173–80. https://doi.org/10.1016/S0378-7753(99)00484-X

[17] Miansari Me, Sedighi K, Amidpour M, Alizadeh E, Miansari Mo. Experimental and thermodynamic approach on proton exchange membrane fuel cell performance. *Journal of Power Sources* 2009;190:356–61. https://doi.org/10.1016/j.jpowsour.2009.01.082

[18] Herez A, El Hage H, Lemenand T, Ramadan M, Khaled M. Review on photovoltaic/thermal hybrid solar collectors: Classifications, applications and new systems. *Solar Energy* 2020;207:1321–47. https://doi.org/10.1016/j.solener.2020.07.062

[19] Nualboonrueng T, Tuenpusa P, Ueda Y, Akisawa A. Field experiments of PV-thermal collectors for residential application in Bangkok. *Energies* 2012;5, 1229-1244. https://doi.org/10.3390/en5041229

[20] U.S. Environmental Protection Agency, Combined Heat and Power Partnership. *Technology Characterization: Reciprocating Internal Combustion Engines.* Washington, D.C.: United States Environmental Protection Agency; 2015.

[21] Al-Sulaiman FA, Hamdullahpur F, Dincer I. Trigeneration: A comprehensive review based on prime movers. *International Journal of Energy Research* 2011;35:233–58. https://doi.org/10.1002/er.1687

[22] Environmental Protection Agency, Combined Heat and Power Partnership. *Technology characterization: Combustion turbines.* Washington, D.C.: United States Environmental Protection Agency; 2015.

[23] U.S. Environmental Protection Agency, Combined Heat and Power Partnership. *Technology characterization: Fuel cells.* Washington, D.C.: United States Environmental Protection Agency; 2015.

[24] Cheng Z, Guo Z, Tan Z, Yang J, Wang Q. Waste heat recovery from high-temperature solid granular materials: Energy challenges and opportunities. *Renewable and Sustainable Energy Reviews* 2019;116:109428. https://doi.org/10.1016/j.rser.2019.109428

[25] AGICO Cement International Engineering. Clinker Grate Cooler, n.d. http://www.cementmillequipment.com/cooling/clinker-grate-cooler.html

[26] Kawasaki G. *Kawasaki delivers a new waste heat recovery power generation system to taiheiyo cement: This is the first delivery of Kawasaki's new high efficiency waste heat recovery.* Tokyo: Kawasaki Heavy Industries; 2022. https://global.kawasaki.com/en/corp/newsroom/news/detail/?f=20221031_0812

[27] Institute for industrial productivity, International Finance Corporation. Waste heat recovery for the cement sector: Market and supplier analysis; n.d. chrome-extension://efaidnbmnnnibpcajpcglclefindmkaj/https://www.ifc.org/wps/wcm/connect/f0394a25-3645-4765-8291-ea33d9f09594/IFC+Waste+Heat+Recovery+Report.pdf?MOD=AJPERES&CVID=kqgTRfZ13%20November

[28] International Finance Corporation. Waste heat recovery for the cement sector. 2014. https://www.ifc.org/wps/wcm/connect/topics_ext_content/ifc_external_corporate_site/sustainability-at-ifc/publications/report_waste_heat_recovery_for_the_cement_sector_market_and_supplier_analysis

[29] Ma S, Wang J. Experiment and numerical simulation of sinter cooling in annular cooler. *IOP Conference Series Materials Science and Engineering* 2019;612. https://doi.org/10.1088/1757-899X/612/3/032151.

[30] Fan X, Wong G, Gan M, Chen X, Yu Z, Ji Z. Establishment of refined sintering flue gas recirculation patterns for gas pollutant reduction and waste heat recycling. *Journal of Cleaner Production* 2019;235:1549–58. https://doi.org/10.1016/j.jclepro.2019.07.003

[31] Xu Q, Wang K, Zou Z, Zhong L, Akkurt N, Feng J, et al. A new type of two-supply, one-return, triple pipe-structured heat loss model based on a low temperature district heating system. *Energy* 2021;218:119569. https://doi.org/10.1016/j.energy.2020.119569

[32] Chen W, Yin X, Ma D. A bottom-up analysis of China's iron and steel industrial energy consumption and CO2 emissions. *Applied Energy* 2014;136:1174–83. https://doi.org/10.1016/j.apenergy.2014.06.002

[33] Jiu-Ju C. Process of waste heat recovery and utilization for sinter in vertical tank. *China Metallurgy* 2012; 22(1):6–11 (in Chinese).

[34] Sun K, Tseng C-T, Shan-Hill WD, Shieh S-S, Jang S-S, Kang J-L. Model predictive control for improving waste heat recovery in coke dry quenching processes. *Energy* 2015;80:275–83. https://doi.org/10.1016/j.energy.2014.11.070

[35] Zhang S, Wen Z, Xing Y, Liu X, Zhang H, Xiong Y. Experimental study on gas-solid heat transfer characteristics for the vertical waste heat recovery using the inverse problem method. *International Journal of Photoenergy* 2022;2022:4053105. https://doi.org/10.1155/2022/4053105

[36] Qin S, Chang S. Modeling, thermodynamic and techno-economic analysis of coke production process with waste heat recovery. *Energy* 2017;141:435–50. https://doi.org/10.1016/j.energy.2017.09.105

[37] Bisio G, Rubatto G. Energy saving and some environment improvements in coke-oven plants. *Energy* 2000;25:247–65. https://doi.org/10.1016/S0360-5442(99)00066-3

[38] Zhang H, Wang H, Zhu X, Qiu Y-J, Li K, Chen R. A review of waste heat recovery technologies towards molten slag in steel industry. *Applied Energy* 2013;112:956–66. https://doi.org/10.1016/j.apenergy.2013.02.019

[39] Das B, Prakash S, Reddy PSR, Misra VN. An overview of utilization of slag and sludge from steel industries. *Resources, Conservation and Recycling* 2007;50:40–57. https://doi.org/10.1016/j.resconrec.2006.05.008

[40] Sun Y, Zhang Z, Liu L, Wang X. Heat recovery from high temperature slags: A review of chemical methods. *Energies* 2015;8:1917–35. https://doi.org/10.3390/en8031917

[41] Brough D, Jouhara H. The aluminium industry: A review on state-of-the-art technologies, environmental impacts and possibilities for waste heat recovery. *International Journal of Thermofluids* 2020;1–2:100007. https://doi.org/10.1016/j.ijft.2019.100007

[42] Yu M, Gudjonsdottir MS, Valdimarsson P, Saevarsdottir G. Waste heat recovery from aluminum production. In: Sun Z, Wang C, Guillen DP, Neelameggham NR, Zhang L, Howarter JA, et al., editors. *Energy technology 2018*, Cham: Springer International Publishing; 2018, pp. 165–78.

[43] Ladam Y, Solheim A, Segatz M, Lorentsen O-A. Heat recovery from aluminium reduction cells. In: Lindsay SJ, editor. *Light metals 2011*, Cham: Springer International Publishing; 2016, pp. 393–8. https://doi.org/10.1007/978-3-319-48160-9_70

[44] Obara S, Tanaka R. Waste heat recovery system for nuclear power plants using the gas hydrate heat cycle. *Applied Energy* 2021; 292:116667. https://doi.org/10.1016/j.apenergy.2021.116667

[45] world nuclear association. Plans for new reactors worldwide, 2022.https://world-nuclear.org/information-library/current-and-future-generation/plans-for-new-reactors-worldwide.aspx

[46] Safa H. Heat recovery from nuclear power plants. *International Journal of Electrical Power & Energy Systems* 2012;42:553–9. https://doi.org/10.1016/j.ijepes.2012.04.052

[47] Electrical Engineering Info. Nuclear Power Station, 2023. https://www.electricalengineeringinfo.com/2014/12/nuclear-power-station-or-nuclear-power-plant.html

[48] Iturralde J, Gomez deArteche M, Aguirre P, Barcena J, López Pérez S, Ubieta E. Radiant waste heat recovery from steelmaking and glass industry. *E3S Web of Conferences* 2019;116:00029. https://doi.org/10.1051/e3sconf/201911600029

[49] High-Temperature Applications. Waste heat recovery in process industries, 2022, p. 245–67. https://doi.org/10.1002/9783527830008.ch5

[50] Delpech B, Axcell B, Jouhara H. Experimental investigation of a radiative heat pipe for waste heat recovery in a ceramics kiln. *Energy* 2019; 170:636–51. https://doi.org/10.1016/j.energy.2018.12.133

[51] Lund JW, Boyd TL. Direct utilization of geothermal energy 2015 worldwide review. *Geothermics* 2016;60:66–93. https://doi.org/10.1016/j.geothermics.2015.11.004

[52] Huddlestone-Holmes C, Hayward J. The potential of geothermal energy, 2011. https://www.researchgate.net/publication/228591316_The_potential_of_geothermal_energy#fullTextFileContent

[53] BP. Energy outlook: 2022. Available at: https://www.bp.com/content/dam/bp/business-sites/en/global/corporate/pdfs/energy-economics/energy-outlook/bp-energy-outlook-2022.pdf

[54] Kanbur BB, Xiang L, Dubey S, Choo FH, Duan F. Cold utilization systems of LNG: A review. *Renewable and Sustainable Energy Reviews* 2017;79:1171–88. https://doi.org/10.1016/j.rser.2017.05.161

[55] Mokhatab S, Mak JY, Valappil J, Wood DA. *Handbook of liquefied natural gas.* Houston, TX: Gulf Professional Publishing; 2013.

[56] BP. Statistical review of world energy, 2021, 70th edition. Available at: https://www.bp.com/content/dam/bp/business-sites/en/global/corporate/pdfs/energy-economics/statistical-review/bp-stats-review-2021-full-report.pdf

# 2 Waste Thermal Energy Utilization Technologies

## 2.1 RECOVERING WASTE HEAT ENERGY

Recent research has revealed that 72% of primary energy could be saved globally if the produced waste heat can be effectively harnessed [1]. The grade of the waste heat must be judiciously considered during the employment of heat recovery technologies. That is because different technologies achieve their optimal heat recovery performance at different temperature ranges. Conventionally, the waste heat energy is categorized into (1) low-temperature range (<100°C), (2) medium-temperature range (100–300°C), and (3) high-temperature range (>300°C) [1]. In addition, the capacities of the heat recovery system are highly dependent on the quantity of the waste heat and consumers' demands. Thermal energy storage systems such as heat exchangers or thermal materials are classified as passive heat recovery technologies. In contrast, heat recovery technologies that transfer heat to other forms of energy are labeled as active technologies [2]. Typically, waste heat is recovered to produce cooling, domestic heating, and electricity simultaneously.

### 2.1.1 Recovering Waste Heat to Generate Electricity

#### 2.1.1.1 Organic Rankine Cycle (ORC)

1. Introduction of ORC

    An organic working fluid with a low boiling point is employed in the Organic Rankine cycle (ORC) to generate electricity by feeding on the thermal energy released from a low-grade waste heat source [3]. The ORC technology was invented by Lucien Bronicki and Harry Zvi Tabor in the late 1950s [4]. Figure 2.1 shows the schematic of an ORC. First, the organic working fluid with a low boiling point is pumped into the evaporator. Then, the evaporated working fluid flows through an expander. After that, the working fluid passes through a condenser and is then condensed. Different working fluids have been employed in ORC, including HCFC123, PF5050, HFC-245fa, HFC245ca, isobutene, and aromatic hydrocarbons [5]. Research has shown that ORCs are able to better recover thermal energy from lower temperature heat sources than the conventional steam Rankine cycle. ORCs are usually installed to harness waste heat from gas turbines, compressor stations, and metal industries. The long payback period is the key impediment to the deployment of ORCs [6].

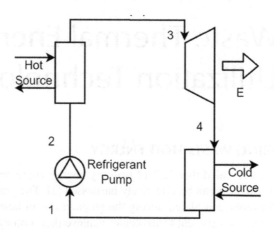

**FIGURE 2.1**  Simplified schematic diagram of ORC [3,7].

The governing equations of ORC are described as:

$$\dot{Q}_{eva} = \dot{m}(h_3 - h_2) \tag{2.1}$$

$$\dot{W}_{exp} = \dot{m}(h_3 - h_4) \tag{2.2}$$

$$\dot{W}_{pump} = \dot{W}_{in} \tag{2.3}$$

$$\eta = \frac{\dot{W}_{exp}}{\dot{Q}_{eva}} \tag{2.4}$$

where $\eta$ represents the efficiency.

2. Experiment case study of ORC

   Liang and Yu [8] investigated the experimental performance of an ORC prototype system. The ORC system comprises an evaporator, a condenser, a diaphragm pump driven by a motor, and an oil-free scroll expander integrated with a single-phase electric generator. In this system, the motor is wired with a variable-frequency inverter, which is used to regulate the flow rate of the working fluid. In this system, the generated electrical power is consumed by the loads (lamps or resistance heaters), which are wired in parallel. The rated powers of the lamp and resistance heater are 400 W and 750 W, respectively. A 7.2 L tank, located at the outlet of the condenser, is used for the storage of working fluid R245fa. Specifications of the system are introduced in Table 2.1. Hot water at 90–97°C forms the heat source to drive the system. Experiment results have shown that the maximum thermal efficiency reaches 4.09% when the pressure ratio is 4.75 (Figure 2.2).

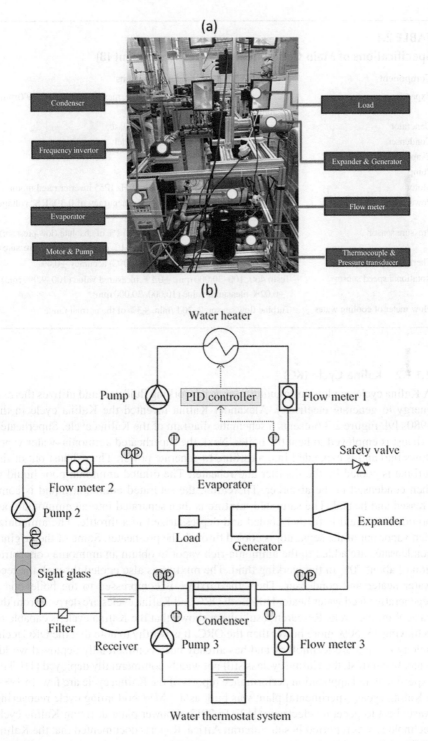

**FIGURE 2.2** (a) Experiment setup of ORC prototype; (b) schematic diagram of the ORC system [8].

**TABLE 2.1**

**Specifications of Main Components and Test Equipment [8]**

| Components | Specifications |
|---|---|
| Expander | Rated power of 1 kW, oil-free, scroll type, maximum speed of 3,600 rpm, volume ratio 3.5 |
| Generator | Connected to expander, 50 Hz, single phase |
| Condenser | Two plate-type heat exchangers in parallel |
| Evaporator | Plate-type heat exchanger |
| Pump | Hydra-cell sealless diaphragm pump |
| Motor | 0.37 kW, six pole 400 V three phase 50 Hz IP55 inverter rated motor |
| Power meter | Measuring error 0.1 (Frequency) Hz, accuracy of 0.3% F.S. voltage and current |
| Pressure sensor | PX319-500A5V, error band (0–500 Psig) 1% of absolute (low pressure) |
|  | PX419-750AV, error band (0–750 Psig) 1% of absolute (high pressure) |
| Thermocouple | K type, 310 stainless steel sheath (1,100°C) accuracy ±0.4% |
| Rotational speed sensor | Testo 460, 100–30,000 rpm, ±0.1% measured value (100–9,999 rpm), ±0.02% measured value (10,000–30,000 rpm) |
| Flow meter of cooling water | Turbine flow sensor, 2–30 L/min, ±3% of the normal range |

### 2.1.1.2  Kalina Cycle (KC)

A Kalina cycle employs ammonia/water as its working fluid pair and utilizes thermal energy to generate electricity. Alexander Kalina invented the Kalina cycle in the 1980s [9]. Figure 2.3 presents a schematic diagram of the Kalina cycle. Superheated exhaust is employed to heat the boiler. Next, the superheated ammonia–water vapor flows into and is expanded into a turbine to generate power. The exhaust out of the turbine is cooled by the distiller and reheater. The diluted ammonia-poor liquid is then condensed in the absorber. Thereafter, the saturated ammonia liquid is compressed and heated. The saturated mixture is then separated into an ammonia-poor liquid. The liquid is further cooled and depressurized in a throttle. The ammonia-rich vapor out of the separator is cooled through the pre-heater. Some of the original condensates are added to the ammonia-rich vapor to obtain an ammonia concentration of about 70% in the working fluid. The mixture is also cooled through the feed water heater and condenser. Then, the mixture is compressed to the boiler via a regenerative feed water heater [10]. Both ORC and Kalina cycles are derived from the basic Rankine cycle. Research results have shown that the Kalina cycle is capable of achieving 15–50% more power than the ORC. It is worthy to note that the ORC technology is a mature technology and has already been commercially deployed worldwide. In contrast, the Kalina cycle is still not widely commercially deployed [11]. The experiment and application performance reports of the Kalina cycle are few. In 1989, a Kalina cycle experimental plant was built as a 3 MW bottoming cycle recovering waste heat to generate electricity [12]. In 2012, a power plant utilizing Kalina cycle technology was reported in sub-Saharan Africa. Reports documented that the Kalina cycle can be used to recycle waste heat generated from power plants, producing up to 20% additional electricity [13].

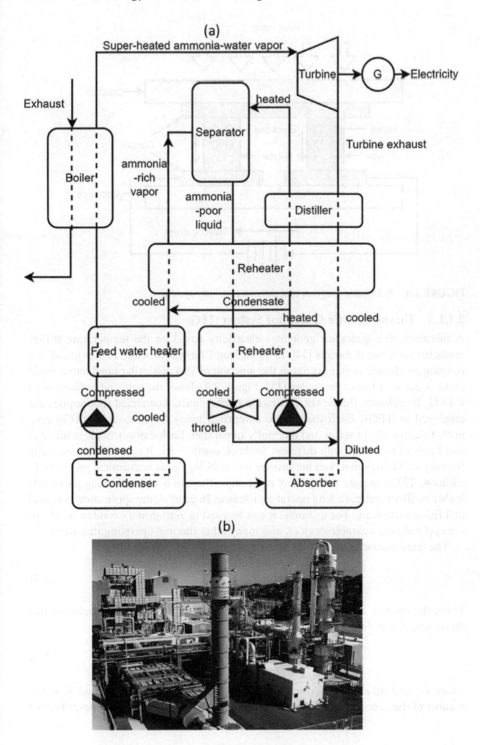

**FIGURE 2.3** (a) Schematic diagram of a simplified Kalina cycle; (b) Kalina cycle technology utilizes waste heat generated by power plants [13].

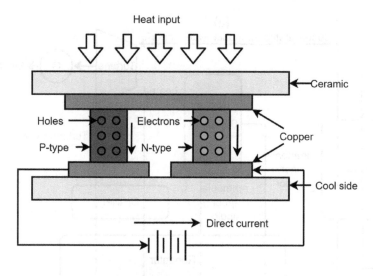

**FIGURE 2.4**  Schematic diagram of a thermoelectricity system.

### 2.1.1.3  Thermoelectric Generator System (TEG)

A thermoelectric generator generates electricity based on the temperature differences between two materials [14]. In 1834, Jean Charles Athanase Peltier found that running an electric current through the junction of two dissimilar conductors could make it act as a heater or cooler [15]. Figure 2.4 shows the schematic diagram of a TEG. To enhance the power generation performance, connected thermopiles are employed in TEGs. Each thermopile comprises many thermocouples (TCs) connected electrically in series and thermally in parallel. Each thermocouple is made of two kinds of materials with different Seebeck coefficients. It is an environmentally friendly technology that does not involve any working fluids or chemical products. In addition, TEGs operate without any noise since they do not have moving parts. It is highly resilient and has a long operating lifetime. In general, the application potential of TEG is significant. For example, it can be used in zero-gravity conditions, deep-sea applications, wearable devices, and many other thermal operating industries.

The temperature gradient is expressed as:

$$\Delta T = T_{\text{hot}} - T_{\text{cool}} \tag{2.5}$$

Then, the electric voltage generated due to the temperature gradient between two thermocouples is described as:

$$V_{\text{out}} = N\alpha_{ab}\Delta T = N\Delta T(\alpha_a - \alpha_b) \tag{2.6}$$

where $\alpha_a$ and $\alpha_b$ are the Seebeck coefficients of the two materials, and $N$ is the number of the connected thermocouples. The delivered output power is described as:

$$P = V_{out}^2 \frac{R_L}{R_{TEG} + R_L} \tag{2.7}$$

The maximum output power is described as:

$$P_{max} = \frac{V_{out}^2}{4R_{TEG}} \tag{2.8}$$

where $R_L$ and $R_{TEG}$ are the external load and internal resistance, respectively.

Liu et al. [16] introduced heat sink-based TEGs for waste heat recovery. A total of 16 TEG units are installed on the top of the combustion chamber and connected in parallel. Each unit consists of 36 modules connected in series. A total of 576 TEG modules are used for the WHR from the ingot casting process. The mold inside the casting chamber is preheated by fire from the combustion of natural gas for casting a large-scale bronze ingot. The flue gas generated in this preheating process moves upward, and waste heat is recovered by installing the TEG assemblies on the top of the chamber. Each module comprises hot- and cold-side insulators, with 391 electrodes on each side, and 391 pairs of N-P-type semiconductors, comprising the materials of ceramic, copper, and Bi2Te3, respectively. The specifications of the proposed system are listed in Table 2.2 (Figure 2.5).

### TABLE 2.2
### Specifications of the TEG System [16]

| Structures/Material | Parameters and Properties | Unit | Values |
|---|---|---|---|
| Electrical insulator/ ceramic | Thickness | mm | 1 |
| | Area | mm | 60*60 |
| | Thermal conductivity | W/m/K | 25 |
| Electrodes | Thickness | mm | 0.3 |
| | Area | mm | 3.7*1.6 |
| | Thermal conductivity | W/m/K | 387.6 |
| | Seebeck coefficient | μV/K | 14 |
| | Electrical resistivity | Ωm | $1.7*10^{-8}$ |
| N-P semiconductor leg/ Bismuth telluride, $Bi_2Te_3$ | Thickness | mm | 0.8 |
| | Area | mm | 1.5*1.5 |
| | Thermal conductivity | W/m/K | Polynomial |
| | Seebeck coefficient | μV/K | Polynomial |
| | Electrical resistivity | Ωm | Polynomial |

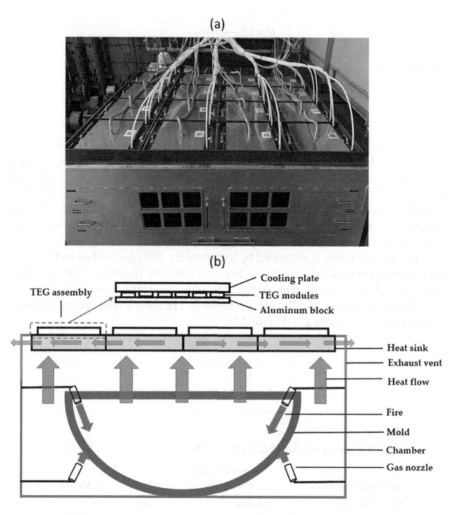

**FIGURE 2.5**   (a) Pictorial view of heat sink-based TEGs; (b) schematic diagram of the TEG system [16].

### 2.1.1.4   Thermophotovoltaic (TPV)

Thermophotovoltaic utilizes heat to produce electricity through the photovoltaic effect. Pierre Aigrain invented thermophotovoltaic between 1960 and 1961 [17]. As illustrated in Figure 2.6, a heat source, radiant emitter, and thermophotovoltaic cell are the main components of TPV. The thermal sources include waste heat, combustion, nuclear energy, and concentrated solar. Compared to a pure solar photovoltaic system, a TPV system is more flexible in utilizing various heat sources. In this regard, TPV is regarded as a promising technology to utilize renewable energy to produce electricity effectively [18].

A band gap is a gap between the valence band and the conduction band of a material. When the energy that is greater than the bandgap energy is provided, photons of the material are emitted and then absorbed by the TPV cell. In this way, electron-hole pairs are generated by the photovoltaic effect and thereby generate electricity. The efficiency of a thermophotovoltaic system is described as:

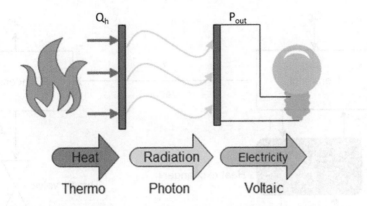

**FIGURE 2.6**   Schematic diagram of TPV.

$$\eta_{TPV} = \frac{P_{out}}{P_{out} + Q_h} = \frac{P_{out}}{P_{inc} + P_{ref}}$$

where $P_{out}$ is the electric power generated by the TPV cell.

TPV is currently not commercially deployed. LaPoint et al. [19] experimentally fabricated and tested TPV cells with efficiencies of more than 40% with the emitter temperatures of 1,900–2,400°C. Thus far, this is the highest reported experimental efficiency of TPV presently. Research predicts that the efficiency of TPV can exceed 50% [20].

## 2.1.2   RECOVERING WASTE HEAT TO PRODUCE COOLING

Absorption and adsorption chillers are well-known technologies that can be employed to recover waste heat to produce cooling. Solid and liquid desiccant dehumidifiers utilize waste heat to remove latent load, thereby improving the energy performance in the air-conditioning processes.

### 2.1.2.1   Absorption Chiller

Absorption chillers are mature commercial heat-driven cooling technology. Edmond Carré developed the first absorption machine in 1850, using water and sulfuric acid [21]. Ferdinand Carré then demonstrated an ammonia/water refrigeration machine in 1859 [21]. A schematic diagram of a single-effect absorption chiller is illustrated in Figure 2.7. Qg, Qe, Qc, and Qa represent the amount of heat transfer in the generator, evaporator, condenser, and absorber, respectively. LiBr/water and ammonia/water are the common working pairs of absorption chillers. A generator, a condenser, an evaporator, and an absorber are the main components of a single-effect absorption chiller. When the hot water heats the working fluids in the generator, water vapor is generated and flows into the condenser (7). The generated strong LiBr solution is released into the absorber (4–6). The condensed LiBr/water in the condenser is further sprayed into the evaporator by utilizing a throttle valve (8–9). The vapor flows into the absorber in a constant manner (10). The strong LiBr absorbs the vapor and is then pumped into the generator (1–3). The chilled water's outlet temperature decreases dramatically since the sprayed vapor absorbs a significant amount of

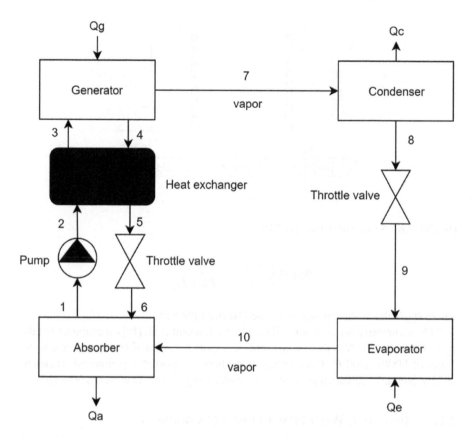

**FIGURE 2.7** Schematic diagram of a single-effect absorption chiller.

heat during the evaporation process. Double-effect and triple-effect absorption chillers can also be designed and employed to recover waste heat from the superheated exhaust. Greater details on the working principles and cycle process of the single/double/triple-effect absorption chillers can be easily obtained from the literature [22]. The inlet and outlet temperature ranges of a single-effect absorption chiller are around 85–100°C and 75–90°C, respectively.

The entire thermal process of the absorption chiller can be described by a set of governing equations. The mass and energy balance within the evaporator is described as:

$$m_9 = m_{10} \tag{2.9}$$

$$Q_e = m_{10}h_{10} - m_9h_9 \tag{2.10}$$

The mass balance equations of the absorber are:

$$m_1 = m_{10} + m_{11} + m_6 \tag{2.11}$$

and

$$x_1 m_1 = x_6 m_6 \tag{2.12}$$

Based on the energy balance within the generator:

$$Q_g = m_4 h_4 + m_7 h_7 - m_3 h_3 \tag{2.13}$$

The energy balance within the solution heat exchanger gives:

$$m_2 h_2 + m_4 h_4 = m_3 h_3 + m_5 h_5 \tag{2.14}$$

The cooling coefficient of performance of the ABC is defined as:

$$COP_{ab} = \frac{Q_e}{Q_g} \tag{2.15}$$

The heat transfer rate of the heat exchanger in the generator is described as:

$$Q_g = U_{gen} A_{gen} \frac{(T_{hwi} - T_4) - (T_{hwo} - T_3)}{\ln\left(\dfrac{T_{hwi} - T_4}{T_{hwo} - T_3}\right)} \tag{2.16}$$

Single-effect, double-effect, and triple-effect absorption chillers are known to be widely deployed to harness waste heat to produce cooling. A commercial LiBr-$H_2O$ single-effect absorption chiller utilizes hot water at 80–100°C to produce chilled water. Both pictorial and schematic perspectives of the absorption chiller are presented in Figure 2.8. The design parameters and experimental performance of the absorption chiller are presented in Tables 2.3 and 2.4. Figure 2.9 presents the varying cooling capacity performance with respect to temperatures of chilled water, cooling water, and hot water.

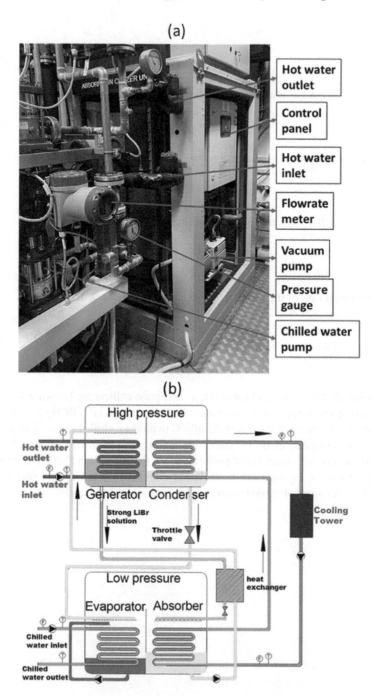

**FIGURE 2.8**  (a) A pictorial view of a single-effect absorption chiller; (b) a schematic view of the single-effect absorption chiller.

## TABLE 2.3
## Nominal Operational Values of the Single-Effect Absorption Chiller

| Parameters | Symbols | Values | Units |
|---|---|---|---|
| Cooling water inlet temperature | $T_{cwi}$ | 30 | °C |
| Hot water inlet temperature | $T_{hwi}$ | 90 | °C |
| Chilled water outlet temperature | $T_{chwo}$ | 7 | °C |
| Cooling water flow rate | $\dot{m}_{cw}$ | 2 | kg/s |
| Chilled water flow rate | $\dot{m}_{chw}$ | 0.83 | kg/s |
| Hot water flow rate | $\dot{m}_{hw}$ | 0.61 | kg/s |
| Rated cooling capacity | $Q_{rcc}$ | 18 | kW |

## TABLE 2.4
## State Parameters of the ABC with $Q_{rcc}$ Equaling to 37.8 kW

| Points | H (kJ/kg) | $\dot{m}$ (kg/s) | %LiBr (X) | P (kPa) | T (°C) |
|---|---|---|---|---|---|
| 1 | 82.09 | 0.2924 | 54.65 | 0.93 | 35 |
| 2 | 85.77 | 0.2924 | 54.65 | 4.80 | 35 |
| 3 | 139.13 | 0.2924 | 54.65 | 4.80 | 54 |
| 4 | 179.24 | 0.2782 | 57.43 | 4.80 | 78 |
| 5 | 123.16 | 0.2782 | 57.43 | 4.80 | 51 |
| 6 | 123.16 | 0.2782 | 57.43 | 0.93 | 45 |
| 7 | 2639.66 | 0.0142 | – | 4.80 | 73 |
| 8 | 162.10 | 0.0142 | – | 4.80 | 39 |
| 9 | 162.10 | 0.0142 | – | 0.93 | 6 |
| 10 | 2511.75 | 0.0142 | – | 0.93 | 6 |

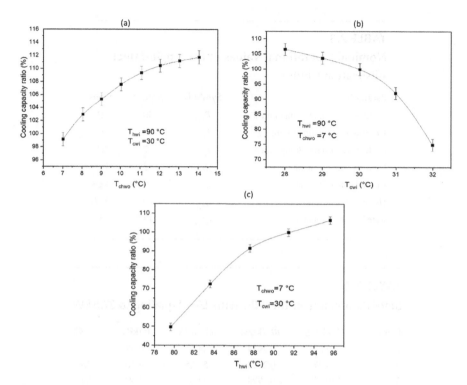

**FIGURE 2.9** Cooling capacity of a single-effect absorption chiller under different operating temperatures.

## 2.1.2.2 Adsorption Chiller

An adsorption chiller is a promising technology that feeds on waste heat to simultaneously produce potable and chilled water [23]. Michael Faraday invented the adsorption chiller in 1821 [24]. A schematic diagram of a four-bed two-evaporator adsorption chiller is presented in Figure 2.10. The main components of the adsorption chillers are a condenser, evaporator, and chambers with desiccant-coated heat exchangers. A spraying mechanism is incorporated into the evaporators to enhance the vapor generation performance. A U-tube is installed to connect the condenser and evaporator so that the condensate water is able to flow back to the evaporator. The hot water circulates through the beds while the cooling water enters the pipes, which go through the beds, evaporators, and condenser. The hot water and cooling water are supplied by a heated water tank and a cooling tower, respectively. During the adsorption process, beds 3 and 4 are connected to the low-pressure evaporator and high-pressure evaporator, respectively. Consequently, the vapor is able to enter the beds from the evaporators. In contrast, beds 1 and 2 are connected to the condenser so that the vapor can flow into the condenser during the desorption process. No vapor is permitted to flow from evaporators to beds or from the beds to the condenser during the switching process because the connecting valves are closed. As depicted in Figure 2.10, each operation cycle comprises an adsorption process, a pre-heating process, a desorption process, and a pre-cooling process. In stage A, the hot water flows through beds 1 and 2. The valves between the beds and the condenser are opened.

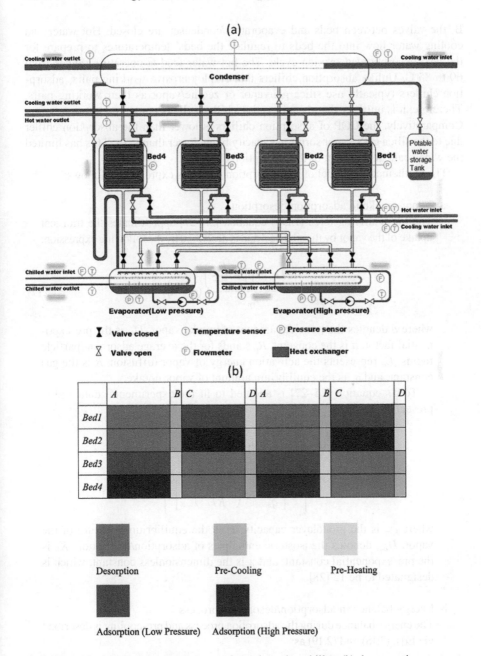

**FIGURE 2.10** (a) A schematic diagram of an adsorption chiller; (b) the operation strategy of four-bed adsorption chiller [24].

Therefore, the vapor is desorbed from the beds to the condenser. Concurrently, beds 3 and 4 are connected to the respective low-pressure evaporator and high-pressure evaporator. The cooling water that is pumped from the cooling tower flows through beds 3 and 4. Accordingly, the vapor is absorbed from the evaporator to the beds. Stage B represents the switching process (pre-cooling/heating process). During stage

B, the valves between beds and evaporators/condenser are closed. Hot water and cooling water flow into the beds to regulate the beds' temperatures to prepare for the next desorption/adsorption cycle. The hot water feed temperature ranges from 60 to 85°C. Unlike absorption chillers that employ erosive working pairs, adsorption chillers typically use silica gel/vapor or zeolite/vapor as their working pairs. Therefore, it is unlikely that the adsorption chiller will encounter any erosion issues. Comparatively, the COP of adsorption chillers is lower than the absorption chiller due to the silica gel's vapor sorption capacity being lower than LiBr. This has limited the wide-scale deployment of the adsorption chillers.

The mathematical model of the adsorption chiller is expressed as follows:

a. Vapor uptake in adsorption/desorption

The linear driving force (LDF) equation [25–27] approximates the transient uptake of the vapor by the silica gel and is described by the following expression:

$$\frac{dq}{dt} = \frac{15 D_{so} \exp\left(-\dfrac{E_a}{RT}\right)}{R_p^2}(q^* - q) \qquad (2.17)$$

where $q$ denotes the transient amount of uptake vapor, $D_{so}$ is the pre-exponential factor, $a$ is the constant, $R_p$ stands for the average adsorbent particle radius, $E_a$ represents the activation energy of vapor diffusion, $R$ is the gas constant, and $q^*$ is the equilibrium amount of vapor uptake.

Tóth's equation [25–27] is adopted to fit the experimental data and is presented as:

$$q^* = \frac{K_0 \exp\left(\dfrac{\Delta H_{ads}}{RT}\right) P_e}{\left\{1 + \left[\dfrac{K_0}{q_m} \exp\left(\dfrac{\Delta H_{ads}}{RT}\right) P_e\right]^\gamma\right\}^{\frac{1}{\gamma}}} \qquad (2.18)$$

where $q_m$ is the monolayer capacity, $P_e$ is the equilibrium pressure of the vapor, $H_{ads}$ denotes the isosteric enthalpies of adsorption/desorption, $K_0$ is the pre-exponential constant, and $\gamma$ is the dimensionless constant, which is designated to be 12 [28].

b. Energy balance in adsorption/desorption process

The energy balance during the adsorption process and pre-cooling is described via Eqs. (2.18) and (2.19) as:

$$(c_{p,sg} M_{sg} + c_{p,hex} M_{hex}) \frac{dT_{ads}}{dt} = -k * M_{sg} \Delta H_{ads} \frac{dq_{ads}}{dt} + \dot{m}_{cw} c_{p,w} (T_{cw,i} - T_{cw,o}) \qquad (2.19)$$

$$\frac{T_{cw,o} - T_{ads}}{T_{cw,i} - T_{ads}} = \exp\left(-\frac{U_{cool} A_{hex}}{\dot{m}_{cw} c_{p,w}}\right) \qquad (2.20)$$

where

$$k = \begin{cases} 1, & \text{adsorption process} \\ 0, & \text{pre-cooling process} \end{cases} \quad (2.21)$$

The energy balance during the desorption and pre-heating process can be represented as:

$$(c_{p,sg}M_{sg} + c_{p,hex}M_{hex})\frac{dT_{des}}{dt} = -i * M_{sg}\Delta H_{ads}\frac{dq_{des}}{dt} + \dot{m}_{hw}c_{p,w}(T_{hw,i} - T_{hw,o}) \quad (2.22)$$

$$\frac{T_{hw,o} - T_{des}}{T_{hw,i} - T_{des}} = \exp\left(-\frac{U_{heat}A_{hex}}{\dot{m}_{hw}c_{p,w}}\right) \quad (2.23)$$

where

$$i = \begin{cases} 1, & \text{adsorption process} \\ 0, & \text{pre-cooling process} \end{cases} \quad (2.24)$$

c. Energy balance in evaporator
The energy balance equations during the evaporating process are described as:

$$(c_{p,w}M_{ref,eva} + c_{p,hex}M_{hex,eva})\frac{dT_{eva}}{dt} = -[h_g(P_{eva},T_{ads}) - h_l(T_{eva})]$$
$$(2.25)$$
$$M_{sg}\frac{dq_{ads}}{dt} + \dot{m}_{chw}c_{p,w}(T_{chw,i} - T_{chw,o})$$

$$\frac{T_{chw,o} - T_{eva}}{T_{chw,i} - T_{eva}} = \exp\left(-\frac{U_{eva}A_{eva}}{\dot{m}_{chw}c_{p,w}}\right) \quad (2.26)$$

d. Energy balance in condenser
The energy balance within the condenser can be described by Eqs. (2.27) and (2.28) as:

$$(c_{p,w}M_{ref,con} + c_{p,hex}M_{hex,con})\frac{dT_{con}}{dt} = [h_g(P_{con},T_{des}) - h_l(T_{con})]M_{sg}\frac{dq_{des}}{dt}$$
$$(2.27)$$
$$+ \dot{m}_{cw}c_{p,w}(T_{con,cw,i} - T_{con,cw,o})$$

$$\frac{T_{con,cw,o} - T_{con}}{T_{con,cw,i} - T_{con}} = \exp\left(-\frac{U_{con}A_{con}}{\dot{m}_{cw}c_{p,w}}\right) \quad (2.28)$$

e. System performance equation
The following sets of equations are presented to quantify the chiller's performance:

$$\dot{m}_{vap} = M_{sg}\frac{dq_{des}}{dt} \tag{2.29}$$

where $\dot{m}_{vap}$ represents the rate of vapor from the beds to condenser during desorption process.

$$CC = \frac{\int_0^{t_{cycle}} \dot{m}_{chw}c_{p,w}(T_{chw,i}-T_{chw,o})dt_{eav1} + \int_0^{t_{cycle}} \dot{m}_{chw}c_{p,w}(T_{chw,i}-T_{chw,o})dt_{eav2}}{t_{cycle}} \tag{2.30}$$

$$COP = \frac{\int_0^{t_{cycle}} \dot{m}_{chw}c_{p,w}(T_{chw,i}-T_{chw,o})dt_{eav1} + \int_0^{t_{cycle}} \dot{m}_{chw}c_{p,w}(T_{chw,i}-T_{chw,o})dt_{eav2}}{\int_0^{t_{cycle}} \dot{m}_{hw}c_{p,w}(T_{hw,i}-T_{hw,o})dt} \tag{2.31}$$

$$SDWP = \frac{n\int_0^{t_{cycle}} \dot{m}_{con,cw}c_{p,w}(T_{con,cw,i}-T_{con,cw,o})dt}{h_{fg}(T_{con})M_{sg}} \tag{2.32}$$

where $n$ is 86400.
Adsorption chillers have relatively low efficiencies when compared with electric chillers. As a result, adsorption chiller technologies are not widely deployed. The key challenge in improving the adsorption chiller's efficiency is to develop advanced sorption materials. Figure 2.11 presents a four-bed two-evaporator adsorption chiller

**FIGURE 2.11** A pictorial view of the adsorption chiller.

**TABLE 2.5**
**Nominal Operational Values of the Adsorption Chiller**

| Parameters | Symbols | Values | Units |
|---|---|---|---|
| Hot water flow rate | $\dot{m}_{hw}$ | 0.61 | kg/s |
| Chilled water flow rate | $\dot{m}_{chw}$ | 0.26 | kg/s |
| Cooling water flow rate | $\dot{m}_{cw,\,con}$ | 0.84 | kg/s |
| Cooling water temperature | $T_{cw}$ | 28.4 | °C |
| Operational cycle time | $t_{cycle}$ | 1,270/980 | s |
| Average pressure in high-pressure evaporator | $p_{eva1}$ | $1.56 \times 10^3$ | pa |
| Average pressure in low-pressure evaporator | $p_{eva2}$ | $1.28 \times 10^3$ | pa |
| Average pressure in condenser | $p_{con}$ | $4.23 \times 10^3$ | pa |

with a cooling capacity of 17 kW. The nominal operation parameters are listed in Table 2.5. The transient temperature profile of the system is presented in Figure 2.12. It is observed from Figure 2.13 that both SCP and SDWP improve as the hot water temperature increases. This observation is attributed to the fact that the vapor can be fully regenerated under high temperatures during the desorption process. Consequently, the potential sorption rate driven by the humidity difference is enhanced. In addition, the potential sorption capacity of the desiccant material is also improved during the adsorption process, leading to better SCP and SDWP performance. It is further observed that when the hot water temperature increases, the $COP_{th}$ peaks at 70°C and then depreciate gradually. One plausible reason is that the heat loss of the adsorption chiller becomes excessive at high operating temperatures. The heat loss arises from heating the desiccant material and heat exchangers at a low temperature. Another possible reason is attributed to having a fixed cycle time under a range of varying hot water temperatures. In this case, the adopted cycle time may exceed the required value for the higher hot water temperature. Consequently, a massive supply of hot water leads to a greater amount of energy loss. These observations also indicate that the cycle needs to be shortened at high temperatures to reduce the amount of heat loss. A higher cooling water temperature negatively impacts the chiller's $COP_{th}$, SCP, and SDWP. This is because the equilibrium amount of vapor in the adsorption process improves at low temperatures. In addition, low cooling water temperatures facilitate a greater amount of potable water condensation in the condenser. When the cooling water temperature is 35°C, the SDWP is lowered to 1.5 m³/day/ton. In contrast, the performance of $COP_{th}$, SCP, and SDWP improves with higher evaporator temperatures. This is because the saturated pressure appreciates at higher temperatures. Consequently, more vapor is generated during evaporation, resulting in better cooling effects.

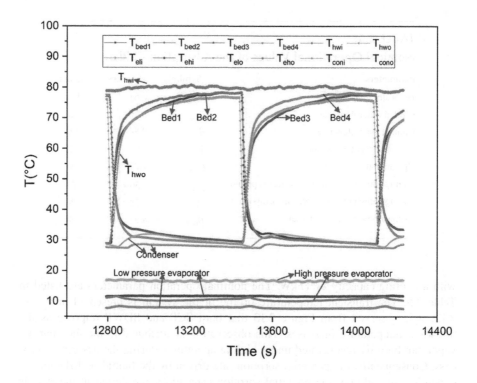

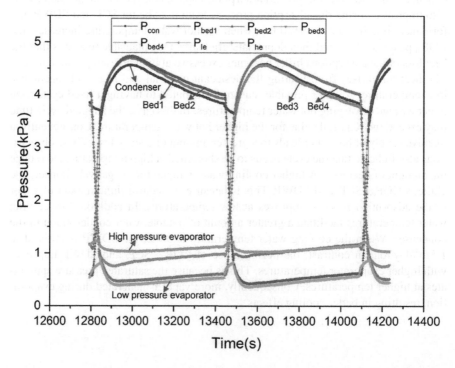

**FIGURE 2.12**   Temperature profile of the four-bed two-evaporator silica gel adsorption chiller.

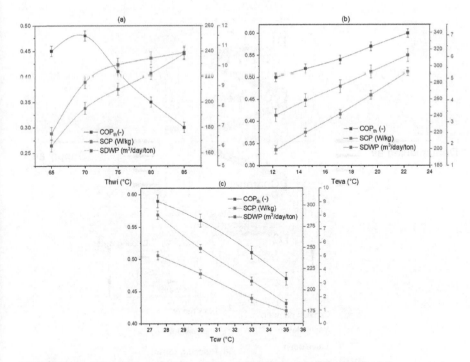

**FIGURE 2.13** Experimental $COP_{th}$, SCP, and SDWP results of the four-bed two-evaporator adsorption chiller considering heat recovery scheme with respect to (a) hot water temperature, (b) evaporator temperature, and (c) cooling water temperature.

## 2.1.2.3 Solid Desiccant Dehumidifier

A thermally driven solid desiccant dehumidifier is a promising technology. It has the capability to decouple sensible and latent cooling in an air-conditioning process, resulting in significant energy performance improvement of cooling systems [29]. Figure 2.14 portrays a schematic diagram of a solid desiccant-coated heat exchanger dehumidifier system. The solid desiccant dehumidifier is installed in two chambers. Each chamber has two heat exchangers coated with silica gel (Type RD). The central working principle leverages the alternation of humidification and dehumidification processes between the desiccant materials at different temperatures. More specifically, during the dehumidification process, the cooling water lowers the temperature of the DCHEs. At the same time, the desiccant material absorbs moisture from the air and releases a certain amount of exothermic heat. Consequently, the humidity of the product air decreases while the temperature increases. In the regeneration process, the hot water heats the desiccant internally and facilitates purging out of the adsorbed moisture carried away by the air stream. In short, the main working principle of the SDCD system is based on the alternation of adsorption and regeneration processes between the desiccant material and the moisture at different temperatures [24]. When chamber 1 operates in the dehumidification mode, the air goes through chamber 1 with valves 1 and 2 opened. Concurrently, chamber 2 functions in the regeneration mode with valves 3 and 4 closed, and the regenerated moisture is exhausted through the damper. Consequently,

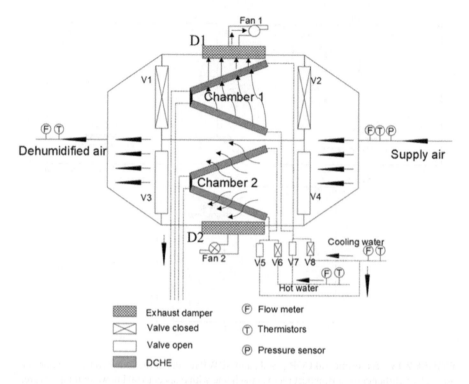

**FIGURE 2.14** Schematic diagram of a solid desiccant-coated heat exchanger dehumidifier [29].

the moisture in the air is removed. The dehumidified air can then be efficiently cooled down with a vapor compression system with lower energy consumption.

Solid desiccant dehumidifiers have shown significant application potential in air-conditioning systems for energy saving. Various desiccant materials are developed to improve the efficiency of solid desiccant dehumidifiers. Vivekh and co-workers have employed different desiccant materials to enhance the sorption capability of DCHEs, including composite superabsorbent polymer and potassium formate [30], superabsorbent polymer, and hygroscopic salt [31]. Wang and co-researchers [32] indicated that several metal–organic frameworks (MOFs) had shown great potential as alternatives to existing desiccant materials, considering their excellent adsorption capability and stable properties. A solid desiccant dehumidifier prototype is illustrated in Figure 2.15. The design and operation parameters are listed in Tables 2.6 and 2.7.

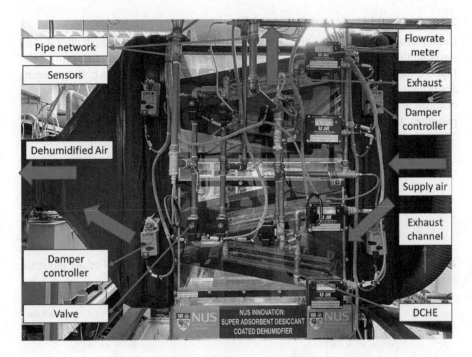

**FIGURE 2.15** A pictorial perspective of the solid desiccant dehumidifier.

### TABLE 2.6
### Geometric Specifications of the
### Fin-Tube Heat Exchanger

| Parameters | Values |
|---|---|
| Length | 700 mm |
| Width | 500 mm |
| Height | 60 mm |
| Fin thickness | 0.12 mm |
| Fin pitch | 1.6 mm |
| Inner diameter of the tube | 10 mm |
| No. of rows | 2 |

### TABLE 2.7
### Operating Conditions of the Dehumidifier System

| Parameters | Units | Design Values | Varying Ranges |
|---|---|---|---|
| Inlet air temperature | °C | 30 | 29.0–32.0 |
| Inlet air humidity | g/kg | 18.2 | 16.0–22.4 |
| Inlet air velocity | m/s | 0.55 | 0.1–4.0 |
| Cooling water temperature | °C | 28 | 27.0–29.0 |

#### 2.1.2.4 Liquid Desiccant Dehumidifier

Liquid desiccant dehumidifiers have long been used for industrial, agricultural, and air-conditioning applications [33]. Figure 2.16 shows the schematic diagram of the liquid desiccant dehumidifier system. Desiccant solutions are filled in both the regenerator and dehumidifier. The working principle is the switching of adsorption and desorption processes between liquid desiccant material and vapor at different temperatures. LiCl, $CaCl_2$, and LiBr are commonly employed liquid desiccant materials. The enthalpy variations of the air and solution can be expressed as:

$$Q_a = m_a(h_{a,in} - h_{a,out}) \tag{2.33}$$

$$Q_s = m_{s,in}(h_{s,out} - h_{s,in}) + m_{de}h_{s,out} \tag{2.34}$$

where $m_a$ and $m_s$ represent the mass flow rate of air and solution. The air moisture removal rate ($m_{de}$) is calculated as $m_{de} = m_a(w_{a,in} - w_{a,out})$.

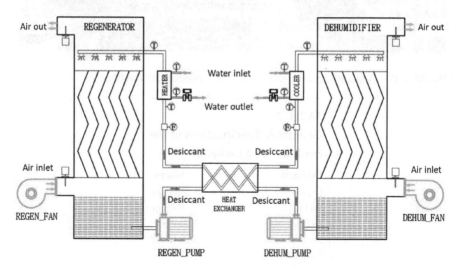

**FIGURE 2.16** Schematic diagram of a liquid desiccant dehumidifier [34].

### 2.1.3 RECOVERING WASTE HEAT TO PRODUCE USEFUL HEATING

#### 2.1.3.1 Heat Pump

Heat pumps are commonly employed to transfer thermal energy from one space to another for heating or cooling purposes. Both electric heat pumps and sorption heat pumps utilize low-grade waste to produce heating [35]. Figure 2.17 portrays a schematic diagram of a vapor-compression heat pump system. It comprises an evaporator, a compressor, a condenser, and an expansion valve. During the vapor compression cycle, the liquid refrigerant absorbs heat in the evaporator (4–1) and releases heat in the condenser (2–3). The heat from the condenser is used to heat air or water.

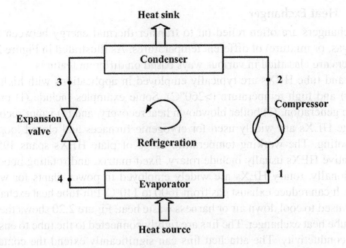

**FIGURE 2.17** Schematic diagram of a vapor compression heat pump system [35].

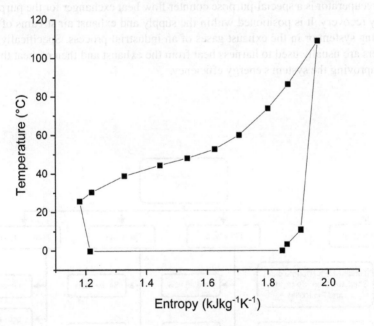

**FIGURE 2.18** System cycle of transcritical $CO_2$ heat pumps [36].

Based on the heat sources, heat pumps are categorized as air sources heat pumps, water sources heat pumps, and ground source heat pumps. The working principles for these heat pumps are similar. The application of heat pumps is determined by the locally available heat sources.

Transcritical $CO_2$ heat pumps have gained substantial research attention due to their abilities to produce much higher-temperature output and excellent carbon emission reduction performance. Figure 2.18 presents the schematic diagram of a transcritical $CO_2$ heat pump. Research showed that transcritical $CO_2$ heat pumps can effectively absorb heat from the air to produce useful hot water at above 60°C.

## 2.1.3.2 Heat Exchanger

Heat exchangers are often relied on to transfer thermal energy between mediums (liquid, gas, or mixture) of different temperatures. As illustrated in Figure 2.19, heat exchangers are classified in various ways based on different features.

Shell and tube HEXs are typically employed in applications with high pressure (>30 bar) and high temperature (>260°C). Some examples include (1) preheating, (2) steam generation, (3) boiler blowdown heat recovery, and (4) vapor recovery system. Plate HEXs are widely used for cryogenic furnaces and closed loop to open water cooling. The working temperature range of plate HEXs spans 195–200°C. Regenerative HEXs usually include rotary, fixed-matrix, and rotating hoods HEXs. Conventionally, rotary HEXs are widely employed in power plants for waste heat recovery. It can reduce exhaust gas from 1600 to 140°C. Fin-tube heat exchangers are typically used to cool down air or harness waste heat. Figure 2.20 shows the diagram of a fin-tube heat exchanger. The fins are tightly connected to the tube to ensure good thermal conductivity. The attached fins can significantly extend the contact areas, which improves heat transfer efficiency.

A recuperator is a special-purpose counter-flow heat exchanger for the purpose of energy recovery. It is positioned within the supply and exhaust air streams of an air handling system or in the exhaust gases of an industrial process. Specifically, recuperators are usually used to harness heat from the exhaust and then preheat the inlet air, improving the system's energy efficiency.

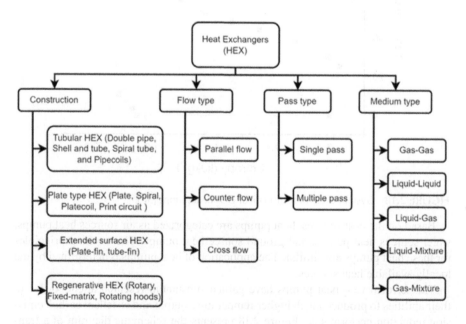

**FIGURE 2.19** Types of heat exchangers.

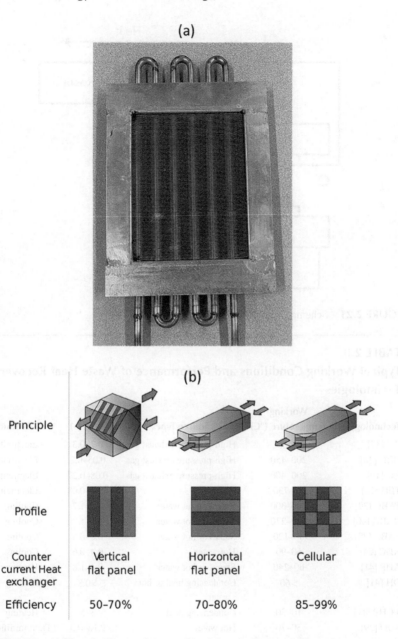

**FIGURE 2.20** (a) Fin-tube heat exchanger; (b) schematic diagram of recuperators.

### 2.1.3.3 Stove Fan

Stove fans are usually placed on the top of a wood-burning stove or an oven. As depicted in Figure 2.21, with the hot and cold sides in position, electricity is generated due to the Peltier effect. Accordingly, the low-torque fan motor is driven. In this way, the heat generated in the stove or oven can be delivered for room heating. A stove or oven will slowly radiate heat into the surrounding room, but it will require

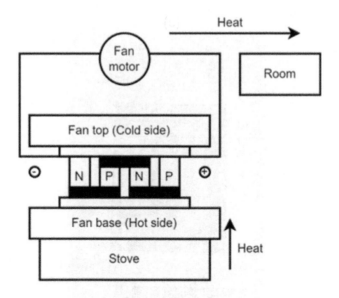

**FIGURE 2.21**    Schematic diagram of stove fan.

**TABLE 2.8**
**Typical Working Conditions and Performance of Waste Heat Recovery Technologies**

| Technology | Working Temperature (°C) | Source Types | COP (–) | Product |
|---|---|---|---|---|
| RC [37] | >400 | High-pressure exhaust gas | 0.2–0.3 | Electricity |
| ORC [38] | 200–450 | High-pressure exhaust gas | 0.05–0.2 | Electricity |
| KC [10] | 200–400 | High-pressure exhaust gas | 0.08–0.27 | Electricity |
| TEG [14] | 50–730 | – | 0.05–0.08 | Electricity |
| TABC [39] | 200–300 | Steam or hot water | 1.4–1.7 | Cooling |
| DABC [40] | 120–270 | Steam or hot water | 1.0–1.2 | Cooling |
| SABC [39] | 80–120 | Steam or hot water | 0.6–0.8 | Cooling |
| ADC [24] | 60–90 | Hot water | 0.4–0.6 | Cooling |
| AHP [41] | 60–240 | Steam or hot water | 0.4–1.8 | Heating |
| DH [41] | >60 | Condensing heat or hot water | >0.8 | Heating |
| VCHP [41] | 10–30 | Condensing heat | 3–5 | Heating |
| SDD [29] | 50–70 | Hot water | 0.15–0.3 | Dehumidified air |

a longer period to heat a room. Another advantage of the stove fan is that it can be placed on the heat sources directly without any battery or wire connection.

   This section briefly introduces several kinds of waste heat recovery technologies that can produce electricity, cooling, and heating. Different waste heat recovery technologies and their working conditions are summarized in Table 2.8.

## 2.2 WASTE COLD ENERGY RECOVERY

### 2.2.1 WASTE COLD ENERGY TO ELECTRICITY

Power generation is one of the most favored LNG cold energy utilization methods since electricity can be conveniently transmitted to users by cable. Based on several Japan's commissioned LNG cold energy utilization facilities, about 76% of them are power generation units, including cryogenic Rankine cycle, natural gas direct expansion (DE), and Rankine/DE combined cycle [42]. Besides the commissioned cycles, other power generation cycles, such as the Brayton cycle and Kalina cycle, are also potential methods to convert the LNG cold energy into electricity.

#### 2.2.1.1 Cryogenic Rankine Cycle

Thus far, the most adopted power generation technology for LNG cold energy utilization is the cryogenic Rankine cycle, including the Rankine cycle integrated inside the combined cycle. The schematic diagram of the cryogenic Rankine cycle for LNG cold energy utilization is illustrated in Figure 2.22. The cryogenic Rankine cycle is similar to the conventional Rankine cycle used in thermal power plants. The working medium goes through four processes, namely, compression, evaporation, expansion, and condensation. A heat source and a heat sink are required to maintain evaporation and condensation. However, the main difference is the operating temperature range. The LNG cold energy plays the role of the heat sink. In other words, the LNG from the storage tank is delivered to cool down the working medium in the condenser. Therefore, the working temperatures in recovering LNG cold energy are significantly lower. Due to the low condensing temperature, water is not a suitable working medium as it may potentially freeze. Fluids with lower boiling and freezing temperatures are more suitable for the cryogenic Rankine cycle, such as the freon or propane hydrocarbon mixture. Propane is one of the ideal working mediums whose boiling and melting points are −42.1°C and −187.6°C, respectively. The heat source of the cryogenic Rankine cycle can be obtained from either seawater or industrial waste heat. The cryogenic Rankine cycle using propane as a working medium was commissioned to recover the LNG cold energy in Japan in 1982. The setup is located in Osaka Gas's Senboku Two receiving terminal, and the power output reached 5840 kW under the LNG regasification rate of 150 t/h [43].

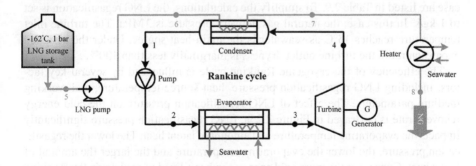

**FIGURE 2.22** Schematic diagram of the cryogenic Rankine cycle.

To evaluate the cold energy utilization efficiency, the total available LNG cold energy is considered as the overall energy input. However, heat energy is not regarded as energy input as the heat source (seawater) is free energy. Therefore, the cold energy recovery rate (CRR) can be defined as:

$$CRR = \frac{W_{net}}{Q_{lng}} = \frac{W_{t,\text{Rankine}} - W_{p,\text{Rankine}} - W_{p,lng}}{m_{lng}(h_8 - h_5)} \tag{2.35}$$

For pumps and turbines, the power consumption and output are, respectively, expressed as:

$$W_{t,\text{Rankine}} = m_{wm}(h_3 - h_4) \tag{2.36}$$

$$W_{p,\text{Rankine}} = m_{wm}(h_2 - h_1) \tag{2.37}$$

$$W_{p,lng} = m_{lng}(h_6 - h_5) \tag{2.38}$$

where $W_{net}$ is the system net power output, $Q_{lng}$ is the available LNG cold energy, $W_{t,\text{ Rankine}}$ is the power output from the Rankine cycle, $W_{p,\text{ Rankine}}$ is the pump power consumption of the Rankine cycle, and $W_{p,\text{ lng}}$ is the LNG pump power consumption.

For the pumps and turbines, the isentropic efficiency is used to describe the performance and is presented as:

$$\eta_p = \frac{h_{p,out,s} - h_{p,in}}{h_{p,out} - h_{p,out}} \tag{2.39}$$

$$\eta_t = \frac{h_{t,in} - h_{t,out}}{h_{t,in} - h_{t,out,s}} \tag{2.40}$$

where $h_{out,s}$ is the ideal component outlet enthalpy, $h_{in}$ is the component inlet enthalpy, $h_{out}$ is the component outlet enthalpy, subscript $p$ is the pump, and subscript $t$ is a turbine.

The detailed thermodynamic parameters of the cryogenic Rankine cycle for a case are listed in Table 2.9. To simplify the calculations, the LNG regasification is set to 1 kg/s. In this case, the natural gas delivery pressure is 3 MPa. The turbine inlet temperature reaches 15°C as seawater is the main heat source. Under the assigned expansion ratio, the turbine outlet dryness is marginally less than 100%.

The efficiency of the cryogenic Rankine cycle is influenced by several key factors, including LNG regasification pressure, heat source temperature, and working medium parameters. The effect of LNG regasification pressure on the cold energy recovery rate is portrayed in Figure 2.23a. LNG regasification pressure significantly impacts the evaporating temperature and amount of latent heat. The lower the regasification pressure, the lower the evaporating temperature and the larger the amount of latent heat. Consequently, more cold energy can be utilized to cool down the Rankine

TABLE 2.9

Detailed Thermodynamic Parameters of the Rankine Cycle with LNG Cold Energy Utilization

| Points | Fluids | m (kg/s) | p (kPa) | T (°C) | h (kJ/kg) | s (kJ/kg/K) | x (%) |
|---|---|---|---|---|---|---|---|
| 1 | Propane | 1.16 | 7 | −88.9 | −0.60 | 0.12 | 0 |
| 2 | Propane | 1.16 | 430 | −88.6 | 0.36 | 0.12 | 0 |
| 3 | Propane | 1.16 | 430 | 15.0 | 602.71 | 2.49 | 100 |
| 4 | Propane | 1.16 | 7 | −88.9 | 469.15 | 2.67 | 99.8 |
| 5 | LNG | 1.00 | 120 | −162.0 | −1.77 | −0.02 | 0 |
| 6 | LNG | 1.00 | 3,000 | −160.4 | 7.93 | 0.01 | 0 |
| 7 | LNG | 1.00 | 3,000 | −93.9 | 551.22 | 3.40 | 100 |
| 8 | LNG | 1.00 | 3,000 | 10 | 844.81 | 4.72 | 100 |
| Overall | CRR | 16.61% | | | | | |

cycle. Hence, the cold energy recovery rate increases as the LNG regasification pressure declines when the working medium's parameters are kept the same, as shown in Figure 2.23b. There is a sharp fall in the efficiency curve. This is attributed to the fact that the LNG evaporating temperature exceeds the condensing temperature of the Rankine cycle. Consequently, the latent heat cannot be utilized, causing a sharp fall in the cold energy recovery rate. Under practical operations, the LNG regasification rate cannot be selected arbitrarily. Instead, it should be determined by the end user's needs and requirements. The required natural gas delivery pressure for different users is listed in Table 2.10. Except for long-distance transportation, the required pressure usually stays below 3 MPa.

Despite the LNG regasification pressure being determined by users, parameter optimization is still possible to be conducted for the cryogenic Rankine cycle to improve its efficiency. The impact of condensing pressure on the cold energy recovery rate is highlighted in Figure 2.24a. To facilitate a fair comparison, the minimum temperature difference ($\Delta T_{min}$) between the working medium and LNG should be kept at a reasonable value, i.e., 5°C. As shown in this figure, although a larger quantity of cold energy is utilized by the condenser under the higher condensing pressure, the power output is lower. The observation can be elucidated from the fact that the cold energy recovery rate deteriorates at elevated condensing pressures when the turbine inlet parameters are maintained. Hence, a good design practice is to let the condensing temperature approach the LNG evaporating temperature. In addition, the minimum temperature difference of 5°C is only a conventional guide in most heat exchanger designs. A smaller minimum temperature difference leads to a lower condensing pressure. However, the required heat transfer area increases as the minimum temperature difference becomes smaller, resulting in a higher capital cost. Therefore, the trade-off between cycle efficiency and capital investment should be optimized through the employment of multi-objective algorithms.

Earlier, it has been highlighted that a lower LNG regasification pressure is beneficial for the cold recovery rate, but the pressure has to be determined based on

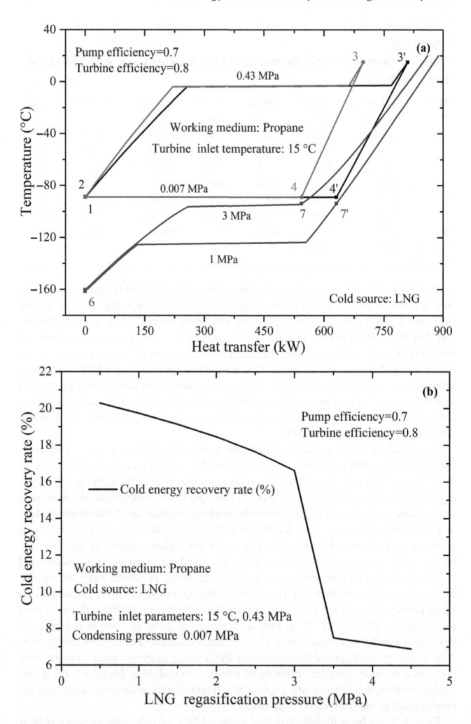

**FIGURE 2.23** Impact of LNG regasification pressure on the cycle performance: (a) T-Q diagram; (b) cold energy recovery rate.

**TABLE 2.10**
**Required Natural Gas Pressure**
**from Different Users**

| Users | Pressure (MPa) |
|---|---|
| Steam power stations | 0.6 |
| Combined cycle stations | 2.5 |
| Local distribution | 3.0 |
| Long-distance distribution | 7.0 |

the users' demand. In addition, lower condensing pressure of the working medium is also good for achieving higher efficiencies. Besides the condensing parameters, there are two other key parameters that need to be judiciously considered, namely, turbine inlet pressure and temperature. The effect of turbine inlet pressure on the cold energy recovery rate is illustrated in Figure 2.25a. As the turbine inlet pressure increases, the cold energy recovery rate is promoted. This is because a larger expansion ratio leads to larger power output. However, the inlet pressure cannot increase indefinitely. The first concern is the limited heat source temperature. Since the LNG cold energy power generation setup is usually deployed inside the receiving terminal, the available heat source is limited. Meanwhile, fluids, such as freon or propane, are usually employed as the working medium for the cryogenic Rankine cycle, which are flammable mediums. Burning external fuel to heat the working medium raises safety issues. Therefore, seawater is usually selected as a safe and sustainable source for the cryogenic Rankine cycle. The average surface seawater temperature spans from 2°C to 35°C [44]. If the operating pressure is too high, the heat source is unable to vaporize the working medium. The second concern is the turbine outlet dryness. The turbine outlet dryness decreases at higher turbine inlet pressures, as depicted in Figure 2.25b. In other words, the working medium will partially condense on the turbine blade when the dryness is less than 100%. However, the turbine outlet dryness should not be less than 88%, or the turbine blade may suffer from erosion issues [45]. All in all, the turbine inlet pressure is limited by the heat source temperature and turbine outlet dryness at a designated condensing pressure.

The impact of the turbine inlet temperature on the cycle performance is shown in Figure 2.26a. When the turbine inlet pressure is maintained constant, a higher turbine inlet temperature leads to larger power output and higher turbine outlet temperature. At the same time, the LNG outlet temperature becomes higher when the turbine outlet temperature increases. This indicates that more LNG cold energy is utilized. The premise is that the minimum temperature difference is controlled at 5°C during the heat transfer process. However, the improvement in power output is marginal. When the turbine inlet temperature exceeds 16°C, the dryness of the working medium at the turbine outlet becomes 100%, as portrayed in Figure 2.26b. This result means that the turbine inlet pressure still has room for improvement. Therefore, the appropriate strategy to enhance the power output is to let the turbine inlet pressure increase along with the improved turbine inlet temperature.

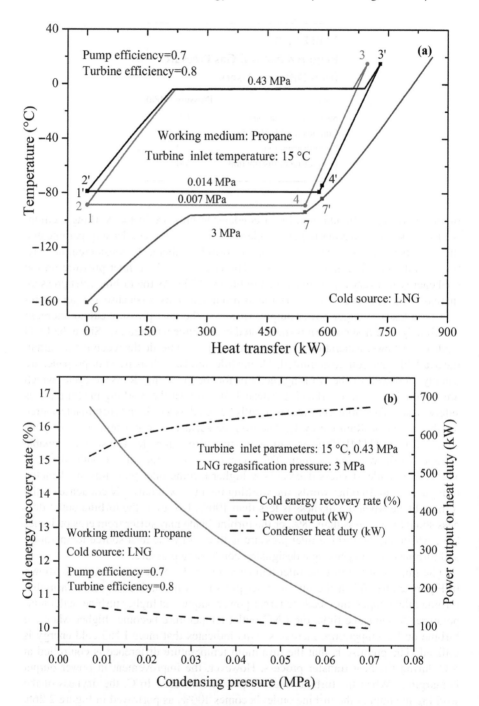

**FIGURE 2.24** Impact of condensing pressure on the cycle performance: (a) T-Q diagram; (b) cold energy recovery rate, power output, and condenser heat duty.

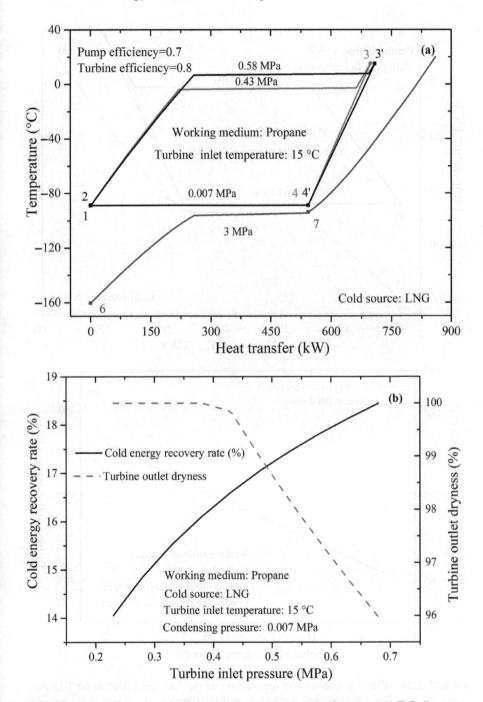

**FIGURE 2.25** Effect of turbine inlet pressure on the cycle performance: (a) T-Q diagram; (b) cold energy recovery rate and turbine outlet dryness.

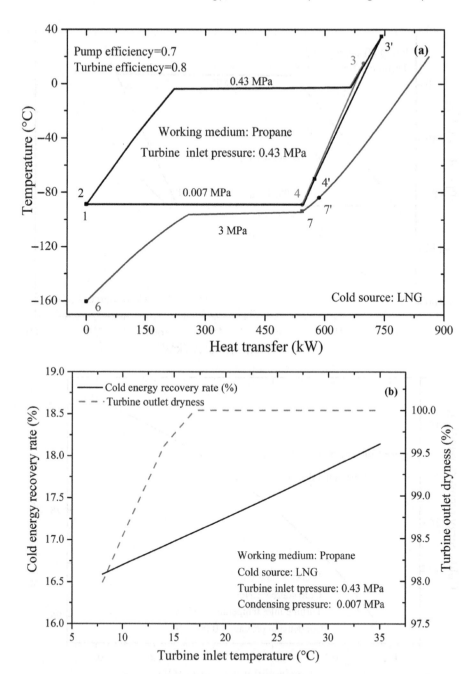

**FIGURE 2.26** Effect of turbine inlet temperature on the cycle performance: (a) T-Q diagram; (b) cold energy recovery rate and turbine outlet dryness.

In sum, adopting good practice in designing an efficient cryogenic Rankine cycle needs to consider many different factors. The LNG regasification pressure is a parameter that is determined by users. A higher turbine inlet pressure leads to a higher expansion ratio, which is beneficial to generate more power output. However, it is worthy to note that the turbine inlet pressure cannot be raised beyond a specific value due to the limitations imposed on the heat source temperature and turbine outlet dryness. The condensing temperature of the Rankine cycle should be designed to close the LNG evaporating temperature. This is because a lower condensing temperature leads to a lower operating pressure, which is beneficial to ensure safe operating conditions and higher cycle efficiency. However, the heat transfer area increases when the minimum heat transfer temperature depreciates. Thus, a multi-objective optimization can be conducted to determine the best possible trade-off outcome among these factors.

## 2.2.1.2  Direct Expansion

Direct expansion (DE) is a special process that utilizes the mechanical exergy (pressure) of LNG instead of pure cold energy. The schematic diagram of the direct expansion process for power generation is illustrated in Figure 2.27. The LNG is first pumped to a pressure that is higher than the distribution network. Next, the high-pressure LNG is vaporized into the gaseous status and drives the direct expansion turbine to produce mechanical work. After the expansion, the natural gas pressure approaches the distribution network pressure. According to commissioned LNG cold energy power generation systems in Japan (total of 15 units), about 1/3 (six units) are purely natural gas direct expansion units. Among these units, natural gas delivery pressure of five units is less than 1 MPa. The LNG cold energy, if left unharvested, is wasted during this process. Thus, the direct expansion process often combines other cycles, such as cryogenic Rankine cycles, to improve the LNG cold utilization efficiency.

Similar to the cryogenic Rankine cycle, the performance of direct expansion is measured from the cold energy side, which is

$$CRR = \frac{W_{net}}{Q_{lng}} = \frac{W_{t,lng} - W_{p,lng}}{m_{lng}(h_5 - h_1)} \tag{2.41}$$

The respective power output and consumption of the LNG turbine and pump are expressed as:

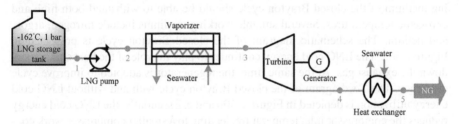

FIGURE 2.27  Schematic diagram of the LNG direct expansion process for power generation.

$$W_{t,lng} = m_{lng}(h_3 - h_4) \tag{2.42}$$

$$W_{p,lng} = m_{lng}(h_2 - h_1) \tag{2.43}$$

In the direct expansion unit, the expansion ratio is the key factor that influences power output. However, the turbine outlet pressure should equal the natural gas delivery pressure, which is determined by users. Therefore, only the turbine inlet pressure can be assigned by the operator. The effect of turbine inlet pressure on the cold energy recovery rate is shown in Figure 2.28a. When the natural gas delivery pressure is the same, the turbine is able to produce more power output as its inlet pressure increases. However, the LNG pump power consumption linearly increases with larger regasification pressure levels, as depicted in Figure 2.28b. Therefore, the cold energy recovery rate does not increase indefinitely with higher LNG regasification pressures. Eventually, there is an optimal regasification pressure that needs to be realized. It is worth noting that the optimal regasification pressure varies with different components' efficiency, including pump and turbine efficiencies. Further, natural gas is a flammable and explosive gas. A higher regasification pressure may jeopardize safe operations.

Another factor that impacts power production is the turbine inlet temperature. As mentioned in the previous sections, the available heat source is limited in the receiving terminal. In addition, LNG is also a flammable medium. Therefore, seawater is also a common heat source for the direct expansion unit. The effect of the turbine's inlet temperature on the cold energy recovery rate is shown in Figure 2.29a. A higher turbine inlet temperature facilitates a larger power output. Nevertheless, the relation between cold energy recovery rate and turbine inlet temperature is linear, which is different from the turbine inlet pressure, as portrayed in Figure 2.29b.

### 2.2.1.3   Closed Brayton Cycle

The traditional Brayton cycle operates in an open system, such as the gas turbine system. These engines incorporate an internal combustor. Hence, the air needs to be replenished from the atmosphere. The fuel gas after the turbine is exhausted into the atmosphere instead of being cooled down. Therefore, the heat exchanger for the heat sink is not necessary. Employing LNG cold energy to cool down the inlet air is an efficient method to reduce compressor work consumption for the Brayton cycle. However, the moisture content of the air can cause freezing issues. Therefore, the closed Brayton cycle is another option to utilize the LNG cold energy. The working medium of the closed Brayton cycle should be able to withstand both high and cryogenic temperatures. Several suitable working mediums include nitrogen, argon, and helium. The schematic diagram of the closed Brayton cycle is presented in Figure 2.30a. The LNG cold energy continues to play the role of the heat sink to cool down the exhaust gas. At the same time, the regenerator is adopted to improve cycle efficiency. The *T-s* diagram of the closed Brayton cycle with and without LNG cold energy utilization is depicted in Figure 2.30b and c. Essentially, the LNG cold energy reduces the compressor inlet temperature, leading to a smaller compressor work consumption when compared to the Brayton cycle operating under ambient temperature. Without the LNG cold energy utilization, the ambient air is employed as the heat

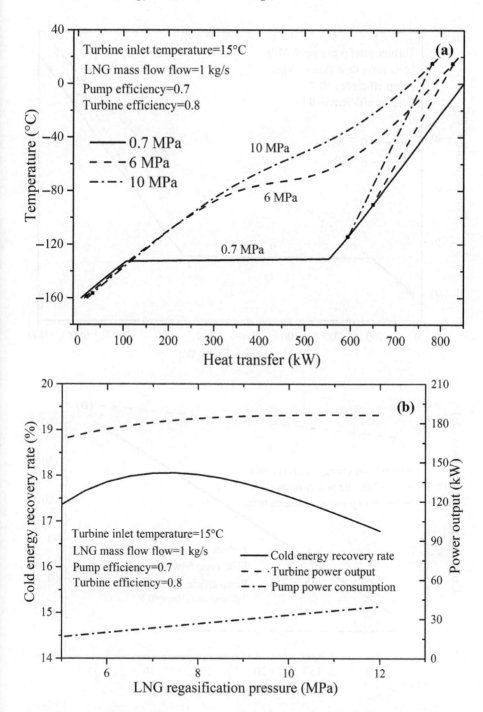

**FIGURE 2.28** Effect of turbine inlet pressure on the cycle performance: (a) T-Q diagram; (b) cold energy recovery rate, turbine power output, and pump power consumption.

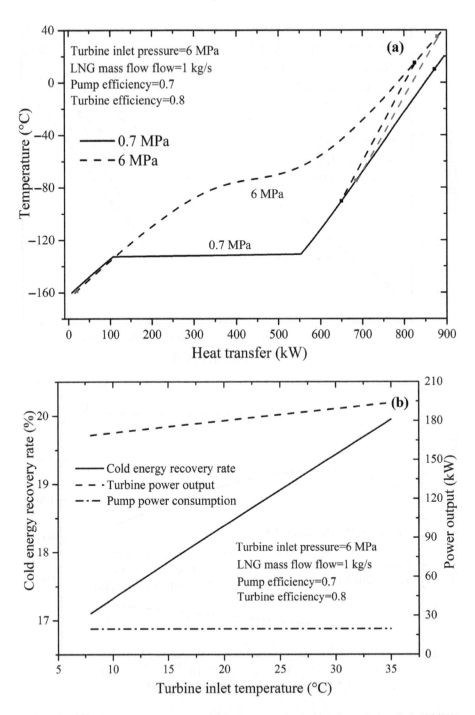

**FIGURE 2.29** Effect of turbine inlet temperature on the cycle performance: (a) T-Q diagram; (b) cold energy recovery rate, turbine power output, and pump power consumption.

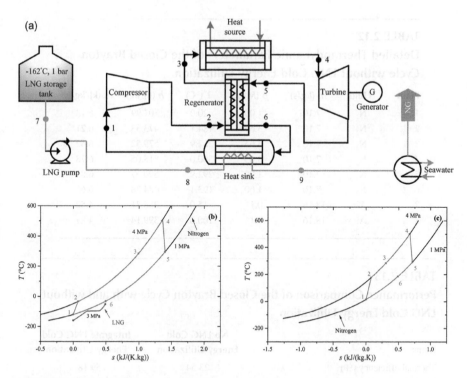

**FIGURE 2.30** Closed Brayton cycle with LNG cold energy utilization: (a) schematic diagram; (b) T-s diagram with LNG cold energy utilization; (c) T-s diagram without LNG cold energy utilization.

**TABLE 2.11**

**Detailed Thermodynamic Parameters of the Closed Brayton Cycle with LNG Cold Energy Utilization**

| Point | Fluid | m (kg/s) | p (MPa) | T (°C) | h (kJ/kg) | s (kJ/kg/K) |
|-------|-------|----------|---------|--------|-----------|-------------|
| 1 | $N_2$ | 7.10 | 1.00 | −120.0 | 150.71 | 5.43 |
| 2 | $N_2$ | 7.10 | 3.65 | −33.9 | 235.64 | 5.50 |
| 3 | $N_2$ | 7.10 | 3.65 | 282.9 | 579.52 | 6.42 |
| 4 | $N_2$ | 7.10 | 3.65 | 500.0 | 818.05 | 6.78 |
| 5 | $N_2$ | 7.10 | 1.00 | 292.9 | 590.57 | 6.83 |
| 6 | $N_2$ | 7.10 | 1.00 | −32.1 | 246.70 | 5.92 |
| 7 | LNG | 1.00 | 0.12 | −162.0 | −1.77 | −0.02 |
| 8 | LNG | 1.00 | 3.00 | −160.4 | 7.93 | 0.01 |
| 9 | LNG | 1.00 | 3.00 | −52.1 | 689.43 | 4.10 |

sink. Accordingly, the compressor inlet temperature is only cooled down to 20°C, which is 5°C higher than the ambient temperature. Thus, the compressor work input is larger under the same compression ratio.

The respective thermodynamic data of the closed Brayton cycle with and without LNG cold energy utilization are summarized in Tables 2.11 and 2.12. The comparison of key

**TABLE 2.12**

**Detailed Thermodynamic Parameters of the Closed Brayton Cycle without LNG Cold Energy Utilization**

| Point | Fluid | m (kg/s) | p (MPa) | T (°C) | h (kJ/kg) | s (kJ/kg/K) |
|---|---|---|---|---|---|---|
| 1 | N$_2$ | 7.10 | 1.00 | 20.0 | 301.99 | 6.13 |
| 2 | N$_2$ | 7.10 | 3.65 | 184.1 | 473.33 | 6.21 |
| 3 | N$_2$ | 7.10 | 3.65 | 282.9 | 579.52 | 6.42 |
| 4 | N$_2$ | 7.10 | 3.65 | 500.0 | 818.05 | 6.78 |
| 5 | N$_2$ | 7.10 | 1.00 | 292.9 | 590.57 | 6.83 |
| 6 | N$_2$ | 7.10 | 1.00 | 193.1 | 484.38 | 6.62 |
| 7 | Air | 18.16 | 0.1 | 15.0 | 286.31 | 4.52 |
| 8 | Air | 18.16 | 0.1 | 100.0 | 399.14 | 4.87 |

**TABLE 2.13**

**Performance Comparison of the Closed Brayton Cycle with and without LNG Cold Energy Utilization**

| Items | No LNG Cold Energy Utilization | Integrate LNG Cold Energy Utilization |
|---|---|---|
| Thermal efficiency (%) | 23.54 | 59.18 |
| Net power output (MW) | 0.398 | 1.0 |
| Compressor efficiency (%) | 80 | 80 |
| Compressor inlet/outlet temperature (°C) | 20.0/184.1 | −120.0/−33.9 |
| Compressor inlet/outlet pressure (MPa) | 1.00/3.65 | 1.00/3.65 |
| Compressor work consumption (MW) | 1.21 | 0.6 |
| Turbine efficiency (%) | 90 | 90 |
| Turbine inlet/outlet temperature (°C) | 500/292.9 | 500/292.9 |
| Turbine inlet/outlet pressure (MPa) | 3.65/1.00 | 3.65/1.00 |
| Turbine work output (MW) | 1.61 | 1.61 |
| Heat energy input (MW) | 1.69 | 1.69 |
| Heat duty of regenerator (MW) | 0.75 | 2.44 |
| LNG mass flow (kg/s) | – | 1 |
| LNG inlet/outlet temperature (°C) | – | −160.4/−52.1 |

parameters is listed in Table 2.13. Under the same pressure ratio, the compressor work consumption is almost half when the LNG cold energy is used to precool the inlet air. Therefore, the thermal efficiency of the closed Brayton cycle with LNG cold energy utilization is more than two times higher than the cycle without LNG cold energy utilization.

## 2.2.2 WASTE COLD ENERGY RECOVERY FOR OTHER PURPOSES

### 2.2.2.1 Air Separation

Air separation is one of the energy-intensive industries. First of all, the atmospheric air is compressed and cooled to a liquid state. Then, the liquified air is sent to the distillation

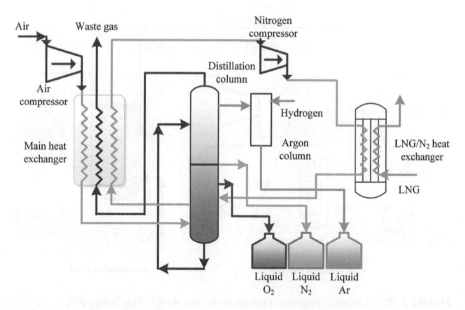

**FIGURE 2.31** Schematic diagram of air separation process using LNG cold energy in Senboku LNG Terminal One [47].

column to separate nitrogen, oxygen, etc. During the liquefication process, a large amount of electricity is consumed to drive the external refrigeration cycle and generate cooling effects. Therefore, replacing the external refrigeration process with LNG cold energy is an excellent option to reduce electricity consumption. The basic principle behind the air separation process involves the use of LNG cold energy as shown in Figure 2.31. This setup was commissioned in Senboku LNG Terminal One of Osaka Gas in 1993 [46]. Since natural gas is a flammable gas, nitrogen is employed as the intermediate working medium. The nitrogen is extracted from the distillation column. This method does not require expanders and external cooling cycles. The operating pressure is 1.96 MPa, and the cycling nitrogen is about 1/5 when compared to a unit with external cooling cycles under the same air separation capacity. The energy consumption for making liquid oxygen is only half compared to the conventional separation process, reaching 2500 kJ/m³. Employing the LNG cold energy to separate air leads to reduced capital investment. Further, the cost used to regasify the LNG can also be reduced at the receiving terminal. The LNG temperature leaving the air separation heat exchanger is about −100°C, which means the cold energy has not been completely exploited. Therefore, users can continue to harvest this remaining cold energy for further utilization.

Another example is a commissioned air separation plant in the vicinity of the LNG receiving terminal of Fos Tonkin in France [47]. The schematic diagram of this plant is displayed in Figure 2.32. The cold energy from the LNG regasification process is partly recovered through the liquefaction process of nitrogen. This process employs a heat exchanger installed at the receiving terminal, where a continuous flow of LNG with a pressure of 70 bar is delivered as the cold source to cool down the gaseous nitrogen. The liquid nitrogen is then sent to the distillation column to separate the oxygen and nitrogen. The specification of the plate-type heat exchanger is listed in Table 2.14. The heat exchanger comprises many thin and slightly separated

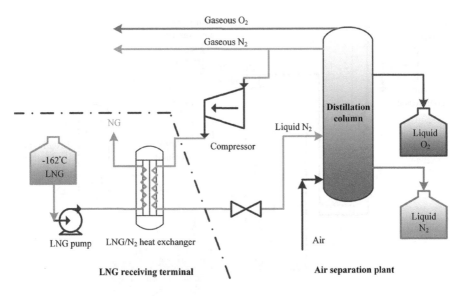

**FIGURE 2.32** Schematic diagram of the air separation plant run by Elengy [48].

**TABLE 2.14**

**The Main Specification of the Heat Exchanger**

| Items | LNG | Nitrogen |
|---|---|---|
| Operating pressure, bar (mini/nominal/max) | 45/72/82 | 33/36/55 |
| Operating temperature, °C (mini/nominal/max) | −160/−160/20 | −110/8/20 (Inflow) |
| Flow rate, m³/h | 100/585/900 (m³ LNG/h) | 84,000/113,000 Nm³/h |
| Nominal operating temperature, °C | | −150 (Outflow) |

plates, which possess very large surface areas and small fluid flow passages. The air separation plant is designed to run separately [48]. In other words, the backup facilities are installed in the air separation plant as contingencies in case of any LNG cold energy disruption.

Instead of power generation, the air separation process is a more efficient way to utilize LNG cold energy. It has been reported that the energy consumption for liquifying LNG is about 2.9 MJ/kg [49]. However, the cold energy release is only about 0.86 MJ/kg when the LNG is heated from -162°C to the ambient temperature of about 15°C. According to previous calculations, the theoretical exergy efficiency of power generation ranges from 12% to 30% due to the limitation of the Rankine cycle. In practice, the exergy efficiency in terms of commissioned plants only lingers around 8%–20% [42]. Thus, it is not economically viable to utilize LNG cold energy to generate electricity. In contrast, if the LNG cold energy can be converted to a low-temperature medium, such as nitrogen at about −150°C in Elengy's air separation plant (Table 2.14), the cold exergy can be efficiently passed and retained.

**FIGURE 2.33** Aerial view of Fos Tonkin terminal and air separation plant [48].

The location of the air separation plant is an important factor that should be carefully considered before adopting LNG cold energy for utilization. The air separation plant has to be close to receiving terminals as the LNG is not safe and economical to undergo long-distance transportation. Figure 2.33 shows the panoramic perspective of the Fos Tonkin terminal and air separation plant. The air separation plant is located in the vicinity of the LNG terminal, and LNG/$N_2$ heat exchanger is deployed inside the receiving terminal to avoid directly sending out the LNG. Nitrogen is regarded as an intermediate working medium to deliver cold energy. On the contrary, air separation should be placed in a convenient transportation network to import raw materials and sell processed products. Therefore, essential infrastructures and policies need to be well planned to support LNG cold energy utilization.

### 2.2.2.2 Liquid $CO_2$ and Dry Ice Production

Liquid $CO_2$ and dry ice are important industrial materials that are widely applied to make sodium carbonate, urea, soda, etc. They are also used in other applications, including welding, sugar refinery, food preservation, farming, and so on. Therefore, the production of liquid $CO_2$ and dry ice is an important industrial activity to meet societal needs. Conventionally, adsorption refrigeration machines are used to cool the raw carbon dioxide gas. Since these machines can only cool the gas to about $-30°C$, the raw gas needs to be compressed to a pressure of about 2 MPa to achieve condensation [48]. If the LNG cold energy can be integrated into the process, the electricity consumption can potentially be reduced by half. The flow chart of the $CO_2$ liquefaction facility with LNG cold energy utilization in Senboku Terminal One of Osaka Gas is illustrated in Figure 2.34. The LNG cold energy is first passed to the intermediary

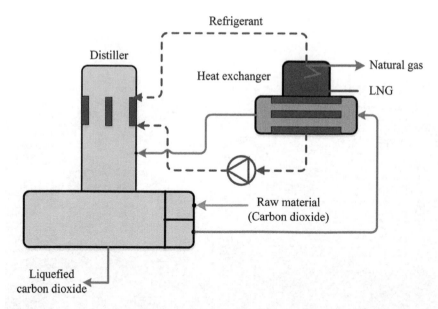

**FIGURE 2.34** Flow chart of $CO_2$ liquefication facility with LNG cold energy utilization [47].

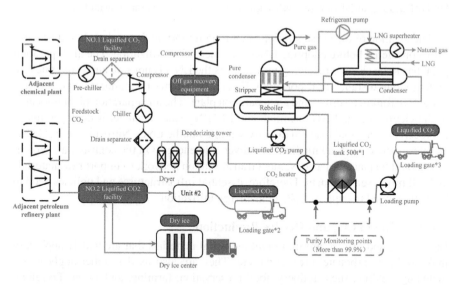

**FIGURE 2.35** Production flow of liquefied carbon dioxide and dry ice [48].

cooling agent (refrigerant). The refrigerant is then sent to the distiller to cool the carbon dioxide. Recently, Osaka Gas company upgraded this technology by incorporating an additional heat exchanger to pre-cool the carbon dioxide. After implementing this configuration upgrade, a 10% electricity reduction for pressurization is realized [50].

Another example is the Kinki Ekitan liquefied $CO_2$ and dry ice production plant in Japan. The process flow of this plant is illustrated in Figure 2.35. The feedstock $CO_2$ is received from an adjacent chemical plant or petroleum refinery plant at 35°C

**TABLE 2.15**
**Production Capacity of Liquified Carbon dioxide and Dry Ice**

| Name | | Specifications |
|---|---|---|
| Facility of liquefied carbon dioxide production | Production capacity | 250 t/day |
| | Liquefied carbon dioxide storage tank | 500 t × 2 |
| | Purity standard of liquefied carbon dioxide | 99.99% or more |
| Facility of dry ice production | Production capacity | 26 t/day × 2 |
| | Storage warehouse for dry ice | Rack type 81 t |

and 0.035 MPa. The $CO_2$ is then pre-chilled, and the dirt it contains is removed. The $CO_2$ is compressed to a state of 170°C and 0.84 MPa by the compressor. The pre-cooled $CO_2$ is dried and deodorized. Thereafter, the $CO_2$ is liquified using the cold energy of the LNG to a temperature of –46°C. Liquified $CO_2$ is stored and shipped on demand. At this stage, the $CO_2$ is –23°C and 2 MPa. For dry ice production, the liquified carbon dioxide is solidified into 25 kg blocks. Sets of 36 blocks are then packed into a flexible container bag to be shipped in a state of –79°C. The production capacity of liquified carbon dioxide and dry ice is shown in Table 2.15.

### 2.2.2.3 Cold Warehouse
The world's first LNG-driven warehouse is located in Yokohama City, Kanagawa Prefecture [48]. LNG cold energy is supplied by the neighboring gas terminal. The operating temperature inside the cold warehouse is maintained in the range of –40 to –60°C. It is used to preserve high-grade marine products, such as tuna, fish eggs, and shrimp. Under Japan's warehouse industry law, warehouses with inside temperatures of below –40°C are termed super freeze warehouses.

The schematic diagram of the LNG-driven refrigeration system is depicted in Figure 2.36. LNG is received from the neighboring gas terminal through a network

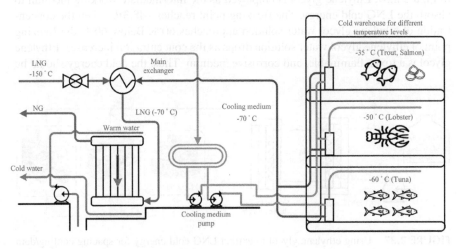

**FIGURE 2.36** Schematic diagram of the LNG-driven refrigeration system [48].

of pipes, reaching –150°C. Refrigerant freon is cooled down by LNG in the main heat exchanger. Next, the LNG flows into the second heat exchanger, where it is heated by warm water and becomes natural gas. Finally, the natural gas is delivered back to the gas terminal and sent back to the city gas network after the calorie adjustment and odor process. The cold refrigerant from the main heat exchanger is temporarily stored in a drum. The refrigerant pump then delivers the refrigerant to the warehouse. In the refrigerator, the cold refrigerant releases its cold energy to the circulating air to sustain a specific cryogenic temperature. Thereafter, the refrigerant from the refrigerator is allowed to flow back to the main heat exchanger.

Two key challenges remain before the wide-scale deployment of LNG cold energy in cold warehouses, including location restriction and low exergy efficiency. The cold warehouse ought to be constructed adjacent to the receiving terminal since the cold energy is not appropriate to experience long-distance transportation. Additionally, the exergy destruction is significant when the high-quality LNG cold energy (–162°C) is directly converted to cooling mediums (–70°C).

### 2.2.2.4 Space Cooling/Data Center Cooling

Space cooling has become the fastest-growing energy demand for buildings in the last three decades [51]. Cooling technologies comprise approximately 30% of buildings' electricity consumption in tropical cities such as Hong Kong [52] and Singapore [53]. The cooling market for buildings is expected to increase continuously and overtake the heating demand in 2070 [54]. Hence, there is to be a huge growing demand for space cooling in the foreseeable future due to the higher requirement for human comfort. Owing to the rapid development of the digital economy, the number of data centers has been increasing quickly in recent years. It has been reported that the energy to support cooling contributes about 30%–50% of electricity consumption in the data center [55]. Harvesting the waste cold energy for space cooling will significantly reduce their electricity consumption and GHG emissions.

The conceptual diagram of spacing cooling by utilizing LNG cold energy is shown in Figure 2.37. Ethylene glycol is employed as the intermediate working medium to absorb the LNG cold energy. The freezing point reaches –48.3°C when the concentration of ethylene glycol water solution approaches 60%. Below 60%, the freezing point of ethylene glycol water solution drops as the concentration increases. Ethylene glycol is a toxic, flammable, and corrosive medium. Thus, the cold energy should be

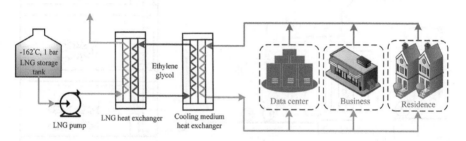

**FIGURE 2.37**   Using ethylene glycol to extract LNG cold energy for spacing cooling/data center cooling.

further transferred to a safe end working medium such as water before being delivered to users.

DaPeng LNG receiving terminal, located in Shenzhen, is the first LNG receiving terminal in China. This terminal was commissioned in 2006, and the designed regasification capacity reaches 800 million tons per year [56]. Initially, all LNG is regasified by ORVs or SCVs in this terminal. Thus far, several ice parks and business conference centers are planned to be built near the receiving terminal to utilize the LNG cold energy. In addition, the cold chain that local fishermen rely upon to store and transport their seafood products will also potentially benefit from the LNG cold energy utilization. In this project, calcium chloride solution is used as the working medium to absorb cold energy from LNG regasification. However, the calcium chloride solution cannot be directly delivered to the users due to its corrosive nature. Accordingly, calcium chloride solution with a temperature of −35°C is first used to make ice, reducing its temperature to -21°C [57]. After that, it is employed to produce chilled water for cooling hotels and stadiums. This project is expected to reduce 20 million kWh electricity consumption and 10,000 tons of carbon dioxide emission.

### 2.2.2.5    Seawater Freeze Desalination

Desalination is a process that removes mineral components from seawater. Commercial desalination technologies are divided into two categories, namely, membrane-based and thermal-based processes. The membrane-based desalination technologies include reverse osmosis (RO), forward osmosis (FO), and electrodialysis (ED). On the contrary, thermal-based processes involve multi-stage flash (MSF), multi-effect distillation (MED), and vapor compression evaporation (VC). Up till 2020, the RO, MSF, and MED contributed 70 %, 18%, and 7% shares of global installed capacity, respectively [58]. However, freeze desalination (FD) is gradually evolving to be an emerging industrial technology and has been used in small- to medium-size seawater desalination plants. Since the latent heat (333.5 kJ/kg) of ice fusion is about seven times smaller than the latent heat (2,256.7 kJ/kg) of water vaporization, freeze desalination is considered a more effective method in contrast to the traditional thermal desalination process. Inexpensive materials and lower maintenance costs arising from the lower operating temperature turn freezing desalination into an economical method.

Now, seawater desalination technology is still an energy-intensive process. It is imperative to provide sustainable and cost-effective cold energy for the freezing process. Since both LNG receiving terminal and seawater desalination plants are located near the coastal area, LNG cold energy can be regarded as a consistent, reliable, and free cold energy provision for seawater desalination. Applying the LNG cold energy to the freeze desalination process not only contributes to the LNG regasification process but also simultaneously produces fresh water. The schematic diagram of the freeze desalination process with LNG cold energy utilization is depicted in Figure 2.38.

The main challenge of freeze desalination technology is the separation of ice from brine. This is the main reason causing the impurity of fresh water. Further, controlling the size of ice crystals formed is also a technical challenge. Constant formation of ice on the heat exchanger surface would significantly deteriorate the heat transfer

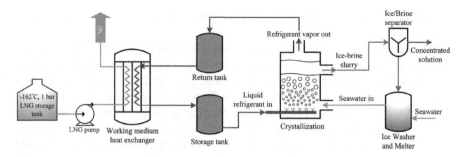

**FIGURE 2.38** Schematic diagram of the freeze desalination process.

coefficient between the cold medium and seawater. Even if the direct contact ice generator is adopted, an inevitable deterioration in heat transfer occurs as the ice fraction increases [59]. Therefore, designing efficient ice/brine separation equipment and well-suited heat exchangers are the keys to enabling LNG cold utilization via the freeze desalination process.

### 2.2.2.6 Cold Energy Storage

Cold energy storage can be regarded as the buffer for LNG cold energy utilization. As the LNG regasification fluctuates according to the users' natural gas demands, the available LNG cold energy does not remain consistent. Thus, the LNG cold energy needs to be transferred to an appropriate form and stored. In addition, the range and flexibility of LNG cold energy utilization can be markedly extended if the cold energy storage medium can be easily transported. The schematic diagram of the cold energy storage using LNG cold energy is shown in Figure 2.39. One of the common cold energy storage systems is to use phase change material (PCM). Usually, the quantity of the latent heat component is larger than the sensible component. For instance, the melting heat of ice is 335 kJ/kg, which is much larger than its sensible heat capacity (4.2 kJ/kg/°C). Hence, transportation efficiency can be improved significantly. Ice is not the only PCM to store cold energy. A detailed introduction to the PCM in cold energy storage will be discussed in Section 3.2.

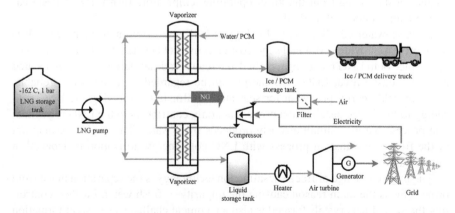

**FIGURE 2.39** Schematic diagram of cold energy storage with LNG cold energy recovery.

Another cold energy storage method is the liquid air energy storage (LAES) system [60]. During the nighttime (off-peak period), the electricity grid has surplus production. Therefore, the liquid air storage module involves consuming the surplus electricity from the power grid and storing it as liquid air in storage tanks. The cold energy from the LNG regasification process is also used to reduce the compression work and liquefy the air. During the daytime (on-peak period), the electricity demand is high. The stored liquid air is heated to become gas again. The gas expands in the air turbine to produce extra electricity. The liquid air storage system plays an essential role in elevating the valleys and shaving the peaks of power grids. The integration of LNG cold energy to a LAES system significantly improves its energy efficiency.

## 2.3 CONCLUSION

This chapter comprehensively introduces various waste thermal energy utilization technologies. Working principles, mathematical models, and application status of these technologies are summarized and analyzed. These technologies are able to utilize thermal energy from hot water, steam, and liquified natural gas, to produce power, useful heating, and useful cooling. In this regard, deploying waste thermal energy utilization technologies can significantly improve energy utilization efficiency, reduce carbon emission, and reduce the operation cost for sustainable energy and environment development.

## REFERENCES

[1] Forman C, Muritala IK, Pardemann R, Meyer B. Estimating the global waste heat potential. *Renewable and Sustainable Energy Reviews* 2016;57:1568–79. https://doi.org/10.1016/j.rser.2015.12.192

[2] Brückner S, Liu S, Miró L, Radspieler M, Cabeza LF, Lävemann E. Industrial waste heat recovery technologies: An economic analysis of heat transformation technologies. *Applied Energy* 2015;151:157–67. https://doi.org/10.1016/j.apenergy.2015.01.147

[3] Ziviani D, Beyene A, Venturini M. Advances and challenges in ORC systems modeling for low grade thermal energy recovery. *Applied Energy* 2014;121:79–95. https://doi.org/10.1016/j.apenergy.2014.01.074

[4] Wikipedia contributors. (2023, June 8). Organic Rankine cycle. In *Wikipedia, The Free Encyclopedia*. Retrieved 13:48, June 19, 2023, from https://en.wikipedia.org/w/index.php?title=Organic_Rankine_cycle&oldid=1159113464

[5] Chen H, Goswami DY, Stefanakos EK. A review of thermodynamic cycles and working fluids for the conversion of low-grade heat. *Renewable and Sustainable Energy Reviews* 2010;14:3059–67. https://doi.org/10.1016/j.rser.2010.07.006

[6] Ancona MA, Bianchi M, Branchini L, De Pascale A, Melino F, Peretto A. Systematic comparison of ORC and s-CO$_2$ combined heat and power plants for energy harvesting in industrial gas turbines. *Energies* 2021;14. https://doi.org/10.3390/en14123402

[7] Shengjun Z, Huaixin W, Tao G. Performance comparison and parametric optimization of subcritical organic rankine cycle (ORC) and transcritical power cycle system for low-temperature geothermal power generation. *Applied Energy* 2011;88:2740–54. https://doi.org/10.1016/j.apenergy.2011.02.034

[8] Liang Y, Yu Z. Experimental investigation of an organic rankine cycle system using an oil-free scroll expander for low grade heat recovery. *International Journal of Green Energy* 2021;18:812–21. https://doi.org/10.1080/15435075.2021.1880915

[9] Karimi M. Thermodynamic analysis of Kalina cycle. *International Journal of Science and Research (IJSR)* ISSN (online): 2319-7064, 2016.

[10] Zhang X, He M, Zhang Y. A review of research on the Kalina cycle. *Renewable and Sustainable Energy Reviews* 2012;16:5309–18. https://doi.org/10.1016/j.rser.2012.05.040

[11] DiPippo R. Second law assessment of binary plants generating power from low-temperature geothermal fluids. *Geothermics* 2004;33:565–86. https://doi.org/10.1016/j.geothermics.2003.10.003

[12] Kalina AI, Leibowitz HM. The Design of a 3MW Kalina Cycle Experimental Plant. *Proceedings of the ASME 1988 International Gas Turbine and Aeroengine Congress and Exposition. Volume 3: Coal, Biomass and Alternative Fuels; Combustion and Fuels; Oil and Gas Applications; Cycle Innovations.* Amsterdam, The Netherlands. June 6–9, 1988; V003T08A004. ASME. https://doi.org/10.1115/88-GT-140

[13] AfricanReview. *AAP carbon granted Kalina cycle license for sub-Saharan Africa.* London: Alain Charles Publishing; 2012. https://www.africanreview.com/energy-a-power/renewables/aap-carbon-granted-kalina-cycle-license-for-sub-saharan-africa

[14] Jaziri N, Boughamoura A, Müller J, Mezghani B, Tounsi F, Ismail M. A comprehensive review of thermoelectric generators: Technologies and common applications. *Energy Reports* 2020; 6:264–87. https://doi.org/10.1016/j.egyr.2019.12.011

[15] Wikipedia. Jean Charles Athanase Peltier. 2022. https://en.wikipedia.org/wiki/Jean_Charles_Athanase_Peltier

[16] Liu J, Shin K-Y, Kim SC. Comparison and parametric analysis of thermoelectric generator system for industrial waste heat recovery with three types of heat sinks: Numerical study. *Energies* 2022;15. https://doi.org/10.3390/en15176320

[17] Robert EN. A brief history of thermophotovoltaic development. *Semiconductor Science and Technology* 2003;18:S141. https://doi.org/10.1088/0268–1242/18/5/301

[18] Burger T, Sempere C, Roy-Layinde B, Lenert A. Present efficiencies and future opportunities in thermophotovoltaics 2020, *Joule* 4(8):1660–1680, ISSN 2542-4351, https://doi.org/10.1016/j.joule.2020.06.021

[19] LaPotin A, Schulte KL, Steiner MA, Buznitsky K, Kelsall CC, Friedman DJ. Thermophotovoltaic efficiency of 40%. *Nature* 2022;604:287–91. https://doi.org/10.1038/s41586-022-04473-y

[20] Omair Z, Scranton G, Pazos-Outón LM, Xiao TP, Steiner MA, Ganapati V. Ultraefficient thermophotovoltaic power conversion by band-edge spectral filtering. *Proceedings of the National Academy of Sciences* 2019;116:15356–61. https://doi.org/10.1073/pnas.1903001116

[21] Fidler JC. A history of refrigeration throughout the world. *International Journal of Refrigeration* 1979;2:249–50. https://doi.org/10.1016/0140-7007(79)90093-8

[22] Jaruwongwittaya T, Chen G. A review: Renewable energy with absorption chillers in Thailand. *Renewable and Sustainable Energy Reviews* 2010;14:1437–44. https://doi.org/10.1016/j.rser.2010.01.016.

[23] Wikipedia. Adsorption refrigeration. 2022. https://en.wikipedia.org/wiki/Adsorption_refrigeration#cite_note-:0-1

[24] Chen WD, Chua KJ. Parameter analysis and energy optimization of a four-bed, two-evaporator adsorption system. *Applied Energy* 2020;265:114842. https://doi.org/10.1016/j.apenergy.2020.114842

[25] Thu K, Ng KC, Saha BB, Chakraborty A, Koyama S. Operational strategy of adsorption desalination systems. *International Journal of Heat and Mass Transfer* 2009;52:1811–6. https://doi.org/10.1016/j.ijheatmasstransfer.2008.10.012

[26] Mitra S, Thu K, Saha BB, Srinivasan K, Dutta P. Modeling study of two-stage, multi-bed air cooled silica gel+water adsorption cooling cum desalination system. *Applied Thermal Engineering* 2017;114:704–12. https://doi.org/10.1016/j.applthermaleng.2016.12.011

[27] Mohammed RH, Mesalhy O, Elsayed ML, Su M, C. Chow L. Revisiting the adsorption equilibrium equations of silica-gel/water for adsorption cooling applications. *International Journal of Refrigeration* 2018;86:40–7. https://doi.org/10.1016/j.ijrefrig.2017.10.038

[28] Sakoda A, Suzuki M. Fundamental study on solar powered adsorption cooling system. *Journal of Chemical Engineering of Japan* 1984;17:52–7.

[29] Chen WD, Vivekh P, Liu MZ, Kumja M, Chua KJ. Energy improvement and performance prediction of desiccant coated dehumidifiers based on dimensional and scaling analysis. *Applied Energy* 2021;303:117571. https://doi.org/10.1016/j.apenergy.2021.117571

[30] Vivekh P, Bui DT, Islam MR, Zaw K, Chua KJ. Experimental performance and energy efficiency investigation of composite superabsorbent polymer and potassium formate coated heat exchangers. *Applied Energy* 2020; 275:115428, ISSN 0306-2619. https://doi.org/10.1016/j.apenergy.2020.115428.

[31] Vivekh P, Islam MR, Chua KJ. Experimental performance evaluation of a composite superabsorbent polymer coated heat exchanger based air dehumidification system. *Applied Energy* 2020;260:114256. https://doi.org/10.1016/j.apenergy.2019.114256

[32] Wang S, Lee JS, Wahiduzzaman M, Park J. A robust large-pore zirconium carboxylate metal–organic framework for energy-efficient water-sorption-driven refrigeration. *Nature Energy* 2018; 3:985–93. https://doi.org/10.1038/s41560-018-0261-6

[33] Mei L, Dai YJ. A technical review on use of liquid-desiccant dehumidification for air-conditioning application. *Renewable Sustainable Energy Review* 2008;12:662–89. https://doi.org/10.1016/j.rser.2006.10.006

[34] Li X, Liu S, Tan KK, Wang Q-G, Cai W-J, Xie L. Dynamic modeling of a liquid desiccant dehumidifier. *Applied Energy* 2016;180:435–45. https://doi.org/10.1016/j.apenergy.2016.07.085

[35] Zhang J, Zhang H-H, He Y-L, Tao W-Q. A comprehensive review on advances and applications of industrial heat pumps based on the practices in China. *Applied Energy* 2016;178:800–25. https://doi.org/10.1016/j.apenergy.2016.06.049

[36] Ye Z, Wang Y, Song Y, Yin X, Cao F. Optimal discharge pressure in transcritical $CO_2$ heat pump water heater with internal heat exchanger based on pinch point analysis. *International Journal of Refrigeration* 2020;118:12–20. https://doi.org/10.1016/j.ijrefrig.2020.06.003

[37] Gewald D, Siokos K, Karellas S, Spliethoff H. Waste heat recovery from a landfill gas-fired power plant. *Renewable and Sustainable Energy Reviews* 2012;16:1779–89. https://doi.org/10.1016/j.rser.2012.01.036

[38] Bianchi M, De Pascale A. Bottoming cycles for electric energy generation: Parametric investigation of available and innovative solutions for the exploitation of low and medium temperature heat sources. *Applied Energy* 2011;88:1500–9. https://doi.org/10.1016/j.apenergy.2010.11.013

[39] Srikhirin P, Aphornratana S, Chungpaibulpatana S. A review of absorption refrigeration technologies. *Renewable and Sustainable Energy Reviews* 2001;5:343–72. https://doi.org/10.1016/S1364-0321(01)00003-X

[40] Deng J, Wang RZ, Han GY. A review of thermally activated cooling technologies for combined cooling, heating and power systems. *Progress in Energy and Combustion Science* 2011; 37:172–203. https://doi.org/10.1016/j.pecs.2010.05.003

[41] Wang X, Jin M, Feng W, Shu G, Tian H, Liang Y. Cascade energy optimization for waste heat recovery in distributed energy systems. *Applied Energy* 2018; 230:679–95. https://doi.org/10.1016/j.apenergy.2018.08.124

[42] Hisazumi Y, Yamasaki Y, Sugiyama S. Proposal for a high efficiency LNG power-generation system utilizing waste heat from the combined cycle. *Applied Energy* 1998; 60:169–82. https://doi.org/10.1016/S0306–2619(98)00034-8

[43] Atienza-Márquez A, Bruno JC, Akisawa A, Coronas A. Performance analysis of a combined cold and power (CCP) system with exergy recovery from LNG-regasification. *Energy* 2019; 183:448–61. https://doi.org/10.1016/j.energy.2019.06.153

[44] NASA Earth obsvervatory. Sea surface temperature. 2022. Available at: https://earthobservatory.nasa.gov/global-maps/MYD28M

[45] Katinić M, Kozak D, Gelo I, Damjanović D. Corrosion fatigue failure of steam turbine moving blades: A case study. *Engineering Failure Analysis* 2019; 106:104136. https://doi.org/10.1016/j.engfailanal.2019.08.002

[46] Yamamoto T, Fujiwara Y, Kitagaki S. Challenges of advanced utilization of LNG Cold in Osaka Gas Senboku LNG Terminals. In: Matsumoto M, Umeda Y, Masui K, Fukushige S, editors. *Design for innovative value towards a sustainable society*, Dordrecht: Springer Netherlands; 2012, pp. 148–53. https://doi.org/10.1007/978-94-007-3010-6_30

[47] Takayuki Y, Yukio F. The accomplishment of 100% utilisation of LNG cold energy-challenges in Osaka Gas Senboku LNG receiving terminals. *Kuala Lumpur World Gas Conference*. Vevey: International Gas Union (IGU); 2012.

[48] Toshiyuki H, Norio H. *Current situation and considerations on LNG cold energy utilization*. Neuilly-sur-Seine: GIIGNL Technical Study Group; 2018.

[49] Łaciak M, Sztekler K, Szurlej A, Włodek T. Possibilities of liquefied natural gas (LNG) use for power generation. *IOP Conference Series: Earth and Environmental Science* 2019;214:012138. https://doi.org/10.1088/1755-1315/214/1/012138

[50] Otsuka T. *Evolution of an LNG terminal*. Amsterdam: Senboku terminal of Osaka Gas; 2006.

[51] IEA. The future of cooling. *Opportunities for efficient air conditioning*. Paris: IEA; 2018. https://www.iea.org/reports/the-future-of-cooling, License: CC BY 4.0.

[52] Lo A, Lau B, Cheng V, Cheung P. *Challenges of district cooling system (DCS) implementation in Hong Kong*. Barcelona: World SB 14; 2014.

[53] Oh SJ, Ng KC, Thu K, Chun W, Chua KJE. Forecasting long-term electricity demand for cooling of Singapore's buildings incorporating an innovative air-conditioning technology. *Energy and Buildings* 2016;127:183–93. https://doi.org/10.1016/j.enbuild.2016.05.073

[54] Balboa-Fernández M, de Simón-Martín M, González-Martínez A, Rosales-Asensio E. Analysis of district heating and cooling systems in Spain. *Energy Reports* 2020;6:532–7. https://doi.org/10.1016/j.egyr.2020.11.202

[55] Jahangir MH, Mokhtari R, Mousavi SA. Performance evaluation and financial analysis of applying hybrid renewable systems in cooling unit of data centers – A case study. *Sustainable Energy Technologies and Assessments* 2021;46:101220. https://doi.org/10.1016/j.seta.2021.101220

[56] CNOOC's Dapeng LNG terminal hits 60 million tons import mark, LNG World News, May 23, 2018. Available at: https://www.offshore-energy.biz/cnoocs-dapeng-lng-terminal-hits-60-million-tons-import-mark/.

[57] Using LNG cold energy to build ice park in DaPeng peninsula. 2018. Available at: Http://WwwSznewsCom/News/Content/Mb/2018-10/25/Content_21170074Htm

[58] Ayaz M, Namazi MA, Ershath MIM, Mansour A, Aggoune el-HM. Sustainable seawater desalination: Current status, environmental implications and future expectations. *Desalination* 2022;540:116022. https://doi.org/10.1016/j.desal.2022.116022

[59] Xie C, Zhang L, Liu Y, Lv Q, Ruan G, Hosseini SS. A direct contact type ice generator for seawater freezing desalination using LNG cold energy. *Desalination* 2018;435:293–300. https://doi.org/10.1016/j.desal.2017.04.002

[60] Qi M, Park J, Kim J, Lee I, Moon I. Advanced integration of LNG regasification power plant with liquid air energy storage: Enhancements in flexibility, safety, and power generation. *Applied Energy* 2020;269:115049. https://doi.org/10.1016/j.apenergy.2020.115049

# 3 Thermal Energy Storage Materials

## 3.1 HEAT ENERGY STORAGE MATERIALS

Heat storage materials are widely employed in various thermal processes to store thermal energy for different applications. It is indispensable to adopt heat storage materials to improve energy utilization efficiency. Based on the storage characteristics, heat storage materials are classified into sensible, latent, and thermochemical heat storage, as presented in Figure 3.1a. Generally, the application processes of thermal energy storage include three processes: thermal energy charging, storage, and discharging. Figure 3.1b summarizes the key factors in developing optimal thermal energy storage materials. The thermal energy storage system should be designed based on specific application requirements, for example, working temperature, volume, and working time. Then, the corresponding materials are selected. The selected storage material should have high storage capacity, excellent heat transfer performance, appreciable durability, and good cyclability performance. In addition, thermal energy storage materials should be cheap and environmentally friendly. The performance and characteristics of different thermal storage materials are briefly presented in Figure 3.2. Based on the specific volume, working temperature, and energy density, the desired thermal energy storage materials can be selected or synthesized to meet the application requirements.

DOI: 10.1201/9781003343387-3

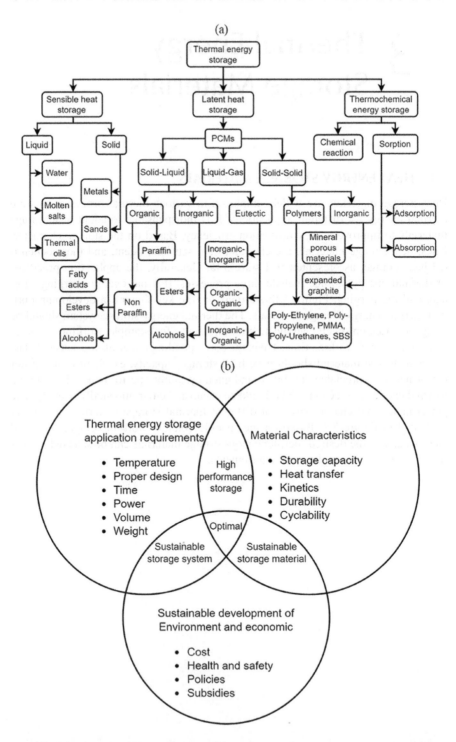

**FIGURE 3.1**   (a) Classification of heat storage materials; (b) key elements to design optimal heat storage system.

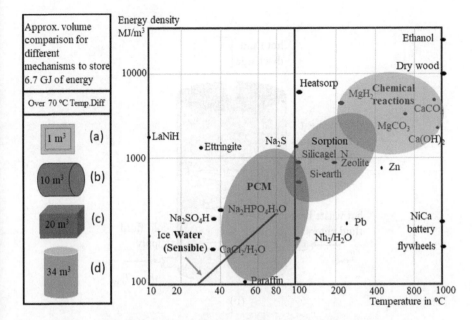

**FIGURE 3.2** Energy density and working temperature of (a) chemical reaction; (b) sorption; (c) PCM; (d) sensible thermal energy storage materials [1].

### 3.1.1 Sensible Heat Storage Materials

Sensible heat storage is realized by increasing the storage material temperature without phase change or chemical reaction. The sensible heat storage materials are usually liquid (water, molten salts, glycol, and oils) or solid (metals, rocks, and sand). The working temperature for rocks and oils is 300–400°C. Comparatively, the molten salts can store heat at 450–850°C [2]. These heat storage materials are typically employed to store thermal energy for domestic hot water and space heating. Sensible thermal energy storage systems are usually integrated with solar or heat pump systems. In this way, thermal energy can be effectively stored. Water is one of the most employed liquid materials that store thermal energy at medium temperatures due to its high volumetric heat capacity and low cost [3]. Figure 3.3 portrays a schematic diagram of a storage tank employing waste as sensible heat storage material. By installing an immersed heat exchanger inside the cylindrical tank, the cold liquid material harnesses thermal energy from the waste hot water. In this way, the waste heat is stored in the heated liquid material.

Al Edhari and Ngo [5] investigated the experimental performance of a thermal energy storage system employing sand, gravel, and pebble rocks as sensible heat storage materials. Results show that the system has a better performance by employing sand as storage material. In addition, the heat dissipation among the storage materials will be reduced with lower porosity.

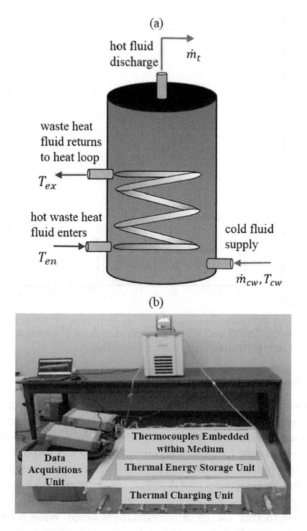

**FIGURE 3.3** (a) A schematic diagram of a storage tank employing fluids as sensible heat storage material [4]; (b) thermal energy storage system using sand, gravel, and pebble rocks as sensible heat storage materials [5].

### 3.1.2 LATENT HEAT STORAGE MATERIALS

In order to conserve energy and utilize sources of heat more effectively, latent heat storage is one of the best methods. The latent heat storage materials are classified into solid-to-liquid, liquid-to-gas, and solid-to-solid phase change materials (PCMs) [6]. These PCMs include organic, inorganic, eutectic, polymer, expanded graphite, and porous mineral materials [6].

#### 3.1.2.1 Organic PCM Materials in Heat Storage

Organic PCMs can absorb and release large quantities of latent heat during phase changes at certain temperatures. Organic PCMs have been widely employed to store heat in various thermal energy processes [7]. Transitions temperatures, heat capacity, thermal conductivity, compositions, the heat of fusion, and long-term characteristics of substances and compounds are key factors that determine the heat storage performance of organic PCMs [8]. Organic PCMs usually include n-alkanes, fatty acids, polyols, polyhydric alcohols, and other organic substances [8]. Figure 3.4 presents some chemical structures of several typical n-alkanes, fatty acids, and polyols.

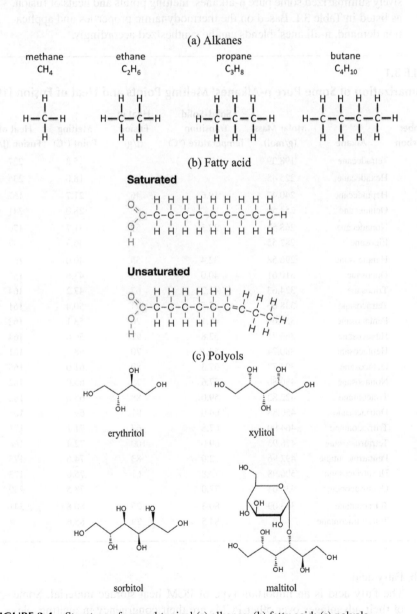

FIGURE 3.4 Structure of several typical (a) alkanes; (b) fatty acid; (c) polyols.

a. Normal alkanes

   Pure n-alkanes and their blends have been widely employed in thermal energy storage. Thermodynamic properties of pure alkanes have been playing key roles in designing suitable heat storage materials. Himran et al. [9] first systematically analyzed the thermodynamic properties of pure alkanes for latent heat storage materials application. Dirand et al. [10] comprehensively summarized some pure n-alkanes' melting points and heats of fusion, as listed in Table 3.1. Based on the thermodynamic properties and application demand, n-alkanes' blends can be synthesized accordingly.

**TABLE 3.1**

**Summarization of Some Pure n-Alkanes' Melting Points and Heat of Fusion [10]**

| Number of Carbon | Alkane | Molar Mass (g/mol) | Solid–Solid Transition Temperature (°C) | Heat of Fusion (J/g) | Melting Point (°C) | Heat of Fusion (J/g) |
|---|---|---|---|---|---|---|
| 14 | Tetradecane | 198.39 | | | 5.8 | 227 |
| 16 | Hexadecane | 225.45 | | | 18.0 | 236 |
| 17 | Heptadecane | 240.47 | 10.6 | 46 | 21.7 | 167 |
| 18 | Octadecane | 254.50 | | | 28.0 | 241 |
| 19 | Nonadecane | 268.53 | 22.3 | 51 | 31.7 | 170 |
| 20 | Eicosane | 282.55 | | | 36.3 | 247 |
| 21 | Heneicosane | 296.58 | 32.4 | 55 | 40.0 | 161 |
| 22 | Docosane | 310.61 | 40.0 | 93 | 43.6 | 157 |
| 23 | Tricosane | 324.63 | 42.2 | 67 | 47.2 | 164 |
| 24 | Tetracosane | 338.66 | 47.5 | 94 | 50.4 | 161 |
| 25 | Pentacosane | 352.69 | 47.1 | 76 | 53.1 | 162 |
| 26 | Hexacosane | 366.71 | 52.8 | 92 | 56.0 | 164 |
| 27 | Heptacosane | 380.74 | 53.1 | 70 | 58.5 | 161 |
| 28 | Octacosane | 394.77 | 57.3 | 89 | 61.0 | 165 |
| 29 | Nonacosane | 480.80 | 57.6 | 75 | 63.0 | 162 |
| 30 | Triacontane | 422.82 | 59.0 | 88 | 65.0 | 162 |
| 32 | Dotriacosane | 450.88 | 64.0 | 91 | 69.3 | 168 |
| 33 | Tritriacontane | 464.90 | 67.8 | 67 | 71.1 | 171 |
| 34 | Tetratriacosane | 478.93 | 69.0 | 100 | 72.4 | 166 |
| 35 | Pentatriacontane | 492.96 | 72.0 | 83 | 74.6 | 175 |
| 36 | Hexatriacosane | 506.98 | 73.8 | 61 | 75.8 | 173 |
| 38 | Octatriacontane | 535.04 | 77.0 | | 78.5 | 249 |
| 40 | Tetracontane | 363.09 | 80.3 | 25 | 80.8 | 241 |
| 44 | Tetratetracontane | | 84.5 | 29 | 85.6 | |

b. Fatty acid

   The fatty acid is an important type of PCM heat storage material. Some of their advantages over other PCMs are their congruency in melting, elevated latent heat of transition, good chemical stability, high specific heat,

### TABLE 3.2
### Melting Point and Latent Heat of Some Fatty Acid Materials

| Name | Melting Point (°C) | Latent Heat (kJ/kg) |
|---|---|---|
| Propyl palmiate | 10 | 186 |
| Isopropyl palmiate | 11 | 100 |
| Oleic acid | 13.5–16.3 | – |
| Isopropyl stearate | 14–19 | 140–142 |
| Caprylic acid | 16 | 148 |
| Butyl stearate | 19 | 140 |
| Dimethyl sabacate | 21 | 120–135 |
| Vinyl stearate | 27–29 | 120 |
| Methyl palmitate | 29 | 120–126 |
| Capric acid | 32 | 152.7 |
| Erucic acid | 33 | – |
| Methyl-12-hydroxy-stearate | 42–43 | 120–126 |
| Lauric acid | 42–44 | 178 |
| Elaidic acid | 47 | 218 |
| Pelargoinc acid | 48 | – |
| Myristic acid | 49–51 | 205 |
| Palmitic acid | 64 | 185.4 |
| Stearic acid | 69 | 202.5 |
| Valporic acid | 120 | – |

nonpolluting supply sources, and nontoxicity [11]. Furthermore, their melting temperature range makes them suitable for passive solar heating [11]. More expensive than paraffins, corrosive, and containing impurities are the main disadvantages of fatty acids [12]. Table 3.2 summarizes the melting point and latent heat of some fatty acid materials. It is noteworthy that fatty acid blends have also been more preferably employed as heat storage materials when compared with pure fatty acid. This is because the blends typically have a more stable long-term thermal storage property and higher heat of fusion.

c. Polyols

As emerging PCMs, polyols are widely employed to harness heat from various thermal processes due to their moderate working temperature range (0–200°C) and significant enthalpies [8]. A polyol is a compound with more than one hydroxyl group [13]. When compared with paraffin as byproducts of fossils, polyols are produced by utilizing sources from plants. In addition, when compared with fatty acids, polyols have demonstrated a better compatible performance with metals. However, it is very expensive to produce pure polyols. Researchers are exploring producing polyols based on carbohydrates hydrogenation to reduce the production cost. The thermal properties of several typical polyols for heat storage are listed in Table 3.3.

**TABLE 3.3**

**Thermal Properties of Several Typical Polyols for Heat Storage [13]**

| Polyols | Melting Temperature (°C) | Heat of Fusion (kJ/kg) |
|---|---|---|
| Polyethylene glycol | 8 | 100 |
| Glycerol | 18 | 199 |
| 1-Tetradecanol | 38 | 205 |
| Ribitol | 102 | 250 |
| Arabinitol | 90 | 230 |
| D-Mannitol | 101 | 256 |
| Catechol | 104 | 207 |
| Thymol | 51.5 | 115 |
| 1-Dodecanol | 11 | 205 |
| Cetyl alcohol | 49 | 141 |
| Palatinitol | 145 | 170 |

### 3.1.2.2   Inorganic PCM Materials in Heat Storage

Inorganic PCMs have been widely employed in building applications for thermal energy management purposes. Similar to other latent heat storage materials, the working principles of inorganic PCMs include absorbing, storing, and releasing thermal energy. By switching the charging and discharging processes, thermal energy management can be realized to achieve energy savings. The latent heat harvesting and releasing processes can be expressed as

$$AB \bullet nH_2O \rightarrow AB \bullet mH_2O + (n-m)H_2O \tag{3.1}$$

$$AB + nH_2O \rightarrow AB \bullet nH_2O \tag{3.2}$$

Inorganic PCMs generally have higher heat storage density, higher operating temperatures, higher thermal conductivity, and lower operating costs wider availability in comparison to organic PCMs [14]. Conversely, corrosivity is the major issue of inorganic PCMs, which may lead to reduced working lifetime and unstable energy storage performance.

The typical inorganic PCMs include salts, salt hydrate, and metallic. The thermal properties of several salts and salt hydrate for heat storage are listed in Table 3.4.

## TABLE 3.4
## Thermal Properties of Several Salts and Salt Hydrate for Heat Storage [14]

| | Melting Temperature (°C) | Heat of Fusion (kJ/kg) |
|---|---|---|
| **Salts** | | |
| $MgCl_2$ | 452 | 714 |
| LiH | 2,678 | 699 |
| $KClO_4$ | 1,253 | 527 |
| KOH | 150 | 380 |
| $KNO_3$ | 266 | 333 |
| $NaNO_3$ | 172 | 307 |
| $LiNO_3$ | 370 | 250 |
| $AlCl_3$ | 280 | 192 |
| **Salt Hydrates** | | |
| $Mg(NO_3)_2 \cdot 6H_2O$ | 150 | 89.3 |
| $MgCl_2 \cdot 6H_2O$ | 169 | 117 |
| $(NH_4)Al(SO_4)_2 \cdot 12H_2O$ | 269 | 95 |
| $Ba(OH)_2 \cdot 8H_2O$ | 266 | 78 |
| $Na_2P_2O_7 \cdot 10H_2O$ | 184 | 70 |
| $CaCl_2 \cdot 6H_2O$ | 140 | 24 |

### 3.1.2.3 Eutectic PCM Materials in Heat Storage

It is noteworthy that a single organic or inorganic PCM cannot cover all the application demands. This is because each PCM possesses a certain melting temperature and enthalpy. Therefore, a type of PCM can only be applied in a given demand condition. Therefore, it is imperative to employ eutectic materials in heat storage. Eutectic materials are a mix of two or more PCMs (organic and/or inorganic). In this way, heat storage materials with desired melting temperature and enthalpy can be customized for various thermal energy storage applications.

The melting temperature and enthalpy of the designed eutectic materials are expressed according to Singh et al. [15]:

$$-\frac{H_a}{T_a}(T_m - T_a) + RT_m \ln(X_a) + G_{a,ex} = 0 \qquad (3.3)$$

$$-\frac{H_b}{T_b}(T_m - T_b) + RT_m \ln(X_b) + G_{b,ex} = 0 \qquad (3.4)$$

where $a$ and $b$ represent two PCM heat storage materials. $T_m$, $T_a$, and $T_b$ denote the melting temperatures of eutectic material, $a$ and $b$, respectively. $H_a$ and $H_b$ are the enthalpies of PCM $a$ and $b$, respectively. $G$ is the excess enthalpy. $X_a$ and $X_b$ represent the mole fraction of $a$ and $b$ in the mixed eutectic material. Table 3.5 lists the properties of a binary eutectic PCM (myristic acid/palmitic acid) with different combinations [16]. More properties of binary eutectic PCMs are listed in Table 3.6 [16].

**TABLE 3.5**
**Properties of a Binary Eutectic PCM (Myristic Acid/Palmitic Acid) [16]**

| Myristic Acid/Palmitic Acid (wt.%) | Melting Temperature (°C) | Enthalpy (J/g) |
|---|---|---|
| 0–100 | 63.08 | 173.69 |
| 10–90 | 59.89 | 172.09 |
| 20–80 | 56.43 | 181 |
| 30–70 | 49.32 | 159.21 |
| 40–60 | 47.95 | 154.29 |
| 50–50 | 47.91 | 153.12 |
| 60–40 | 47.08 | 151.72 |
| 70–30 | 46.73 | 155.43 |
| 80–20 | 49.81 | 154.57 |
| 90–10 | 51.16 | 127.94 |
| 100–0 | 54.7 | 161.37 |

**TABLE 3.6**
**Properties of Various Binary Eutectic PCM Materials [15]**

| Binary Eutectic PCM | Composition (wt.%) | Melting Temperature (°C) | Melting Enthalpy (J/g) |
|---|---|---|---|
| Lauric/capric acid | 35:65 | 18–24 | 140.8 |
| Lauric/stearic acid | 75.5:24.5 | 37.0 | 182.7 |
| Methyl palmitate/methyl stearate | 93–95:7–5 | 23–26.5 | 180 |
| Capric/cetyl alcohol | 70:30 | 22.89 | 144.92 |
| Lauric/myristil alcohol | 40:60 | 21.3 | 151.5 |
| Stearic acid/adipic acid | – | 67.6 | 200.3 |
| Capric acid/stearic acid | – | 21–25 | 154–187 |
| Lauric acid/1-Tetradecanol | 40:60 | 24.33 | 161.45 |
| PEG 2000/PEG10000 | 20:80 | 54 | 185 |
| Lauric/myristic acid | 66:34 | 34.2 | 166.8 |
| Lauric/stearic acid | 75.5:24.5 | 37 | 182.7 |
| Capric acid/lauric acid | – | 18 | 120 |
| Lauric acid/palmitic acid | 69:31 | 35.2 | 166.3 |
| Fatty acids blend/PEG | – | 30–72 | 168–208 |
| Lauric acid/methyl palmitate | 40:60 | 25.6 | 205.4 |
| Lauric acid/nonanoic acid | 96:4 | 41 | 163 |
| Adonitol/erythritol | 70:30 | 87 | 254 |
| 1-Dodecanol/caprylic acid | 30:70 | 6.52 | 171.06 |
| Decanoic acid/tetra decanoic acid | 78:22 | 20.5 | 153 |
| Stearic acid/hexanamide | – | 58 | 176.62 |
| Oxalic acid dehydrate/boric acid | – | 87.3 | 344 |
| Lauric acid/1-dodecanol | – | 17 | 175.3 |
| Lauric acid/hexanediol | 30:70 | 36.92 | 177.11 |

*(Continued)*

**TABLE 3.6 (Continued)**
**Properties of Various Binary Eutectic PCM Materials [15]**

| Binary Eutectic PCM | Composition (wt.%) | Melting Temperature (°C) | Melting Enthalpy (J/g) |
|---|---|---|---|
| N-octadecane/lauric acid | – | 27.89 | 207.6 |
| Stearic acid/n-butyramide | – | 64.41 | 198.38 |
| Tetradecane/hexadecane | – | 1.7–17.9 | 227–246 |
| Capric acid/lauric acid | – | 18.51 | 162.85 |
| Stearic acid/acetamide | – | 64.55 | 193.87 |
| Lauric acid/myristic acid | 66:34 | 34.2 | 166.8 |
| Lauric acid/stearic acid | 75.5:24.5 | 34.16 | 167.30 |
| Capric acid/myristic acid | 72:28 | 18.21 | 148.5 |
| Lauryl alcohol/cetyl alcohol | 80:20 | 20.01 | 191.63 |
| 1,6-Hexanediol/lauric acid | 70:30 | 36.92 | 177.11 |

In addition to binary eutectic PCM materials, various ternary eutectic PCMs have been developed [17]. Some ternary eutectic PCMs have demonstrated appreciable thermal stability after a long cycle test. In order to enhance the thermal conductivity of eutectic PCMs, metallic nanoparticles and carbon-based materials have been attempted to be incorporated into these eutectic PCMs [18]. During the heat absorbing and releasing processes, some PCMs may become liquid. In order to retain the PCMs, polymeric materials are usually employed to develop envelopes around PCMs based on nano/microencapsulation technologies [19]. Additionally, porous or matrix materials are employed to stabilize the shape and form of PCMs in the molten state [20]. More specifically, eutectic PCMs are in molten states when releasing heat. Supporting materials are employed to create a container to prevent the eutectic PCMs from flowing away.

### 3.1.2.4 Polymer Material in Heat Storage

Compared with the organic, inorganic, and eutectic PCMs, polymer PCMs have their distinctive comparative advantages when employed as heat storage materials [21]. For example, polymer PCMs facilitate the stability of the shape and form of heat storage materials. Also, polymer materials can act as encapsulations to reduce the leakage of PCMs. In addition, polymer PCMs can be easily incorporated with desired sized and configurations. Further, polymer PCMs have the highest strength/weight ratio when compared with other types of heat storage PCMs. However, poor thermal conductivity, low heat of fusion, and supercooling are the main disadvantages of polymer PCMs.

Polymer microencapsulation, polymer matrix, and porous polymer framework are three typical methods to develop polymer PCMs. In polymer microencapsulation, phase change material is held in a solid or liquid state while preventing the surrounding medium from interacting with it. In polymer matrices, PCMs are incorporated into different matrixes, which provide stable supporting structures or thermal conducting layers. Comparatively, a porous polymer framework has a higher strength/weight ratio and appreciable insulating properties (Table 3.7).

**TABLE 3.7**
**Properties of Polymer-Based PCMs [21]**

| Methods | Polymer-Based PCM | Melting Temperature (°C) | Latent Heat (J/g) |
|---|---|---|---|
| Polymer microencapsulation | Myristic acid/dicumyl peroxide/poly (methyl methacrylate) | 65.11 | 123 |
| Polymer microencapsulation | n-eicosane/poly (methyl methacrylate) | 35.2 | 84.2 |
| Polymer microencapsulation | Stearic acid/poly (methyl methacrylate) | 60.4 | 92.1 |
| Polymer microencapsulation | Stearic acid/ polycarbonate | 60.0 | 91.4 |
| Polymer microencapsulation | Lauric acid/polystyrene | 43.77 | 167.26 |
| Polymer matrix | Palmitic acid/ polyethylene | 62 | 157.82 |
| Polymer matrix | Palmitic acid/polypyrrole | 59.8 | 160.9 |
| Polymer matrix | Myristic acid/polyaniline | 57.35 | 150.63 |
| Polymer matrix | Paraffin/epoxy | 60.2 | 152.6 |
| Polymer matrix | Hexadecanol/ polyurethane | 50.3 | 229.5 |
| Polymer porous framework | Fatty acid/polyurethane foams | 39.25 | 210.08 |
| Polymer porous framework | Paraffin/polyurethane foams | 54.1 | 210.6 |
| Polymer porous framework | Bio-based coconut fat/ cellulose | 22.63 | 106.17 |

### 3.1.2.5   Expanded Graphite Material in Heat Storage

Expanded graphite is a porous material with excellent adsorption capabilities [22]. Expanded graphite is commonly employed to improve thermal conductivity and reduce the leakage of PCMs due to its high thermal conductivity, cheapness, and low mass density [23]. Fabrication method, concentration, dimension, and packing density are the key factors that affect the thermal conductivity of the expanded graphite-based PCMs [24]. For example, the thermal conductivity of RT44HC can be improved by 20–60 times when the percentage of expanded graphite is increased from 25 to 30 wt% [25].

### 3.1.3   THERMOCHEMICAL HEAT STORAGE MATERIALS

Thermochemical heat storage materials, process design, and reactor have significant effects on the heat storage system performance [26]. Thermochemical heat storage materials can store approximately eight to ten times more heat than conventional heat storage materials and have twice the storage volume of conventional heat storage

[26]. In addition, when compared with other sensible or latent heat storage materials, thermochemical heat storage materials have higher storage density, lower volume requirements, low heat loss, and lower charging temperature [26]. Thermochemical heat storage materials include chemical reactions and sorption heat storage materials.

### 3.1.3.1  Thermochemical Reaction Heat Storage Materials

The reversible chemical reaction processes can be simply described below:

$$A + \Delta H_r \Leftrightarrow B + C \tag{3.5}$$

$$Q_r = n_A \Delta H_r \tag{3.6}$$

where $n_A$ and $\Delta H_r$ are the mol number of reactant/product A and reaction enthalpy, respectively. $Q_r$ is the thermal energy. During the thermal energy charging process, $A$ is the reactant, while $B$ and $C$ are products. During the thermal energy discharging process, $B$ and $C$ are the reactants. During this process, $A$ is produced with heat released. Simply speaking, when an endothermic reaction occurs, heat is stored. And when an exothermic reaction occurs, heat is released. The energy density is typically defined as:

$$D_v = \frac{Q_r}{V_r} \tag{3.7}$$

where $V_r$ is the material volume and $D_v$ represents the volumetric energy density (Table 3.8).

**TABLE 3.8**
**Thermochemical Heat Storage System at Medium and High Temperatures [27]**

| Types | Materials | Reaction Process | Reaction Temperature |
|---|---|---|---|
| Metallic hydrides | $MgH_2$ $CaH_2$ | $MH_n + \Delta H_r \Leftrightarrow M + \dfrac{n}{2}H_2$ | 1,223–1,373 K |
| Carbonates system | $PbCO_3$ $CaCO_3$ | $MCO_{3(s)} + \Delta H_r \Leftrightarrow MO_{(s)} + CO_{2(g)}$ | 973–1,273 K |
| Hydroxides system | $Mg(OH)_2$ $Ca(OH)_2$ | $M(OH)_{2(s)} + \Delta H_r \Leftrightarrow MO_{(s)} + H_2O_{(g)}$ | 523–723 K |
| Redox system | $BaO_2$ $Co_3O_4$ | $M_xO_{y(s)} + \Delta H_r \Leftrightarrow xM_{(s)} + \dfrac{y}{2}O_2$ | 623–1,373 K |
| Ammonia system | $NH_4HSO_4$, $NH_3$ | $NH_4HSO_{4(l)} + \Delta H_r \Leftrightarrow NH_{3(s)} + H_2O_{(g)} + SO_{3(g)}$ <br> $2NH_{3(g)} + \Delta H_r \Leftrightarrow N_{2(g)} + 3H_{2(g)}$ | 673–973 K |
| Organic system | $CH_4 / H_2O$ $CH_4 / CO_2$, $C_6H_{12}$, $SO_3$ | $CH_{4(g)} + H_2O_{(l)} + \Delta H_r \Leftrightarrow CO_{(g)} + 3H_{2(g)}$ <br> $CO_{(g)} + H_2O_{(l)} \Leftrightarrow CO_{2(g)} + H_{2(g)} + \Delta H_r$ <br> $C_6H_{12(g)} + \Delta H_r \Leftrightarrow C_6H_{6(g)} + 3H_{2(g)}$ <br> $SO_3 + \Delta H_r \Leftrightarrow 2SO_{2(g)} + O_{2(g)}$ | 566–1,373 K |

### 3.1.3.2 Sorption Heat Storage Materials

Generally, these materials are classified into adsorption materials (e.g., MOFs, zeolites), absorption materials (e.g., LiBr/H$_2$O), and salts [28]. Excellent cyclic stability and high energy storage density are the main advantages of these sorption thermochemical heat storage materials. Figure 3.5a presents different categories of sorption

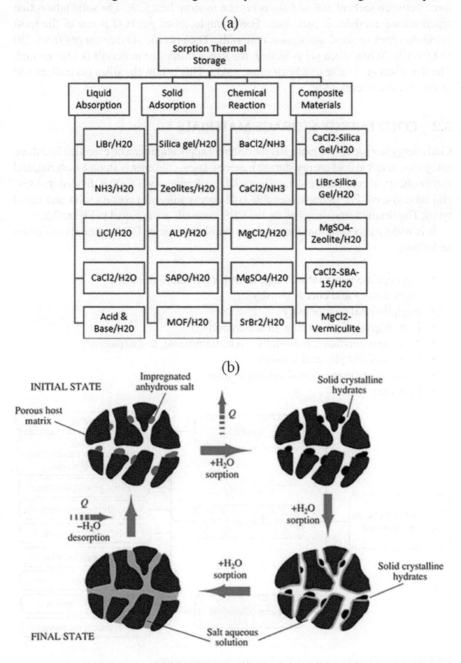

**FIGURE 3.5** (a) Types of sorption thermochemical heat storage; (b) sorption processes with thermal energy charging and discharging processes [28].

thermochemical heat storage materials. Based on the working pairs, the sorption heat storage materials are classified into liquid absorption, solid adsorption, chemical reaction, and composite materials [28].

Figure 3.5b illustrates the sorption processes, including thermal energy charging and discharging. In the sorption thermochemical heat storage process, the binding force between sorbent and sorbate is broken to store heat [29]. The solid adsorption reactions are exothermic processes. For example, silica gel/$H_2O$ is one of the most common types of solid adsorption materials. The specific surface ranges from 750 to 850 $m^2$/g. When silica gel is heated, the vapor inside the materials is regenerated. The dry silica gel is able to adsorb vapor again. This is how the silica gel utilizes and stores thermal energy [30].

## 3.2   COLD ENERGY STORAGE MATERIALS

Cold energy storage (CES) materials for different applications can be classified into three categories, sensible cold energy storage material, latent cold energy storage material, and thermochemical cold energy storage material. Each category is further divided into several subdivisions. For example, sensible cold energy materials contain solid and liquid types. The detailed classifications for the CES materials are depicted in Figure 3.6.

It is widely accepted that the common requirements for CES materials are given as follows:

- Appropriate temperature range according to the application
- High density and energy density
- High thermal conductivity
- Low degree of subcooling (or supercooling)
- No or less corrosive, poisonous, toxic, flammable, and explosive
- Chemical and physical stability
- Small vapor pressure and volume change
- Low cost

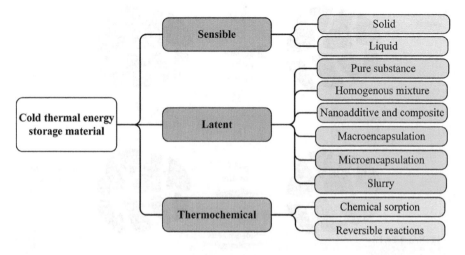

**FIGURE 3.6**   Classification of CES for sub-zero temperatures.

However, no material can meet all the requirements. Therefore, it is a trade-off process to find the most suitable material for the specific application.

### 3.2.1 Sensible Cold Energy Storage Material

Sensible cold energy storage systems are usually simpler and less expensive than latent cold energy storage systems [31]. The most suitable materials for sensible cold energy storage systems are solid, such as rocks, plastics, metals, quartz, concrete, and rock pebbles. Their storage temperature ranges from –170°C to room temperature. These materials are usually integrated with the packed-bed systems, as shown in Figure 3.7a. As solid sensible cold energy storage materials usually present high density and low specific heat, the size and weight are the key aspects to evaluate the feasibility of such projects.

Liquid sensible cold energy storage materials act as both energy storage material and heat transfer fluid (HTF) in an energy storage system. The schematic of a twin-tank liquid cold storage system is shown in Figure 3.7b. Freezing and boiling points are the two key parameters that need to be considered when using such materials. In other words, the working temperature should be kept between the two temperatures to avoid any phase change. In addition, the vapor pressure is better to remain below 1 atm to obtain a safer operation condition. The popular materials that meet these requirements are hydrocarbons and their derivates. Nevertheless, these materials are usually toxic and flammable. Thus, health and safety are the major considerations. Additionally, the shorter chain alcohols and their water solutions are also suitable for liquid sensible cold energy storage, such as methanol, ethanol, and propylene glycol. However, these materials become viscous when the temperature is low. Thus, the operating temperature range should be carefully selected.

### 3.2.2 Latent Cold Energy Storage Material

Latent cold energy storage materials, which are also called phase change materials (PCMs), store energy during materials' phase transition. PCMs can store remarkably more energy than sensible cold energy storage materials. For instance, the fusion energy of water is 335 kJ/kg, while the specific heat is only 4.2 kJ/kg/K. However, the system configuration is also more complex for the phase change energy utilization than the sensible energy utilization [31].

Latent cold energy storage can be implemented by four different processes, including the solid-solid, solid-liquid, solid-gas, and liquid-gas transitions. Solid-liquid transition is the most popular form of PCM. Solid-solid transition releases lower latent energy than the solid-liquid transition as it changes between the polymorphic phases [33]. Solid-gas and liquid-gas transitions contribute the largest amount of cold energy accompanied by a significant volume change during the phase change. Consequently, these materials require vessels that can withstand large volume change or open-air systems. When open-air systems are considered, safe and eco-friendly materials, such as air, nitrogen, and carbon dioxide, should be chosen.

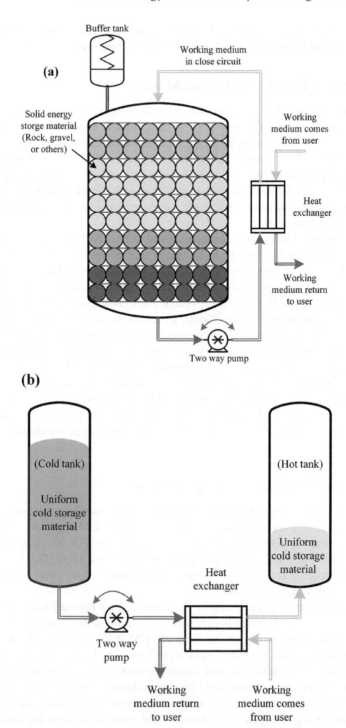

**FIGURE 3.7** Sensible cold energy storage system: (a) packed-bed thermocline, (b) twin tank liquid cold store [32].

### 3.2.2.1 Pure Substance PCM

Pure substance PCMs can be classified into two types, including organic and inorganic. Alkanes (paraffins), alkenones, esters, fatty acids, alcohols, glycols, and polymers are common materials. They belong to carbon and hydrogen structures. Alkanes are the most widely studied type of organic PCM [34]. The thermal properties of some typical alkanes are listed in Table 3.9. They are less expensive, reliable, predictable, non-corrosive, and available for a wide temperature range. The applicable temperature ranges from –200°C to 0°C for cold energy storage purposes. Most organic PCMs have low or no subcooling, but the volume change could be significant during the phase change [35]. Meanwhile, they are usually flammable and sometimes oxidizing.

Inorganic PCMs mainly comprise elements, salts, salt hydrate, and metals. Among them, the most basic options are elements such as nitrogen, hydrogen, helium, oxygen, and neon. Their transition temperatures are below –173°C, which indicates they are suitable for extremely low-temperature storage. Dry ice and liquid carbon dioxide are widely used in food and industrial sectors. Liquid nitrogen is an important cold material in the medical, food, and research sectors. Ammonia has the largest fusion energy among all pure substances collected by the work [36]. However, it is toxic and corrosive.

### 3.2.2.2 Homogenous Mixture PCM

Homogenous mixture PCM is defined as the combination of two or more substances. They are combined in physical form instead of chemical form, and the composite is uniform. The most used homogenous mixture PCM is eutectic material. Eutectic materials can be organic or inorganic. Inorganic eutectic PCMs are inexpensive, with high latent heat and thermal conductivity. However, they are chemically unstable and corrosive. Eutectic water-salt solution is the most widely analyzed and applied PCM for sub-zero temperature ranges [35]. The concentration of the mixture can be adjusted to obtain different phase change temperatures.

Organic homogenous PCMs are also studied by some researchers. A mixture of vegetable oil and water was studied in the literature [37]. It is shown that the freezing point of water has been decreased while maintaining similar thermal properties of water. Vegetable oil can also reduce the subcooling effect. The aqueous alcohol solutions, such as glycol and ethanol, are also potential homogenous PCMs for the storage temperature ranges from –60°C to 0°C [38]. The freezing point of the solution decreases as the concentration increases. However, the latent heat of fusion

**TABLE 3.9**
**Thermal Properties of Paraffins [34]**

| PCM | Formula | Melting Temperature (°C) | Enthalpy of Fusion (kJ/kg) | Liquid Density (kg/m³) |
|---|---|---|---|---|
| n-Nonane | $C_9H_{20}$ | –53.5°C | 120.6 | 720 |
| n-Decane | $C_{10}H_{22}$ | –29.6°C | 201.8 | 730 |
| n-Undecane | $C_{11}H_{24}$ | –25.5°C | 141.9 | 740 |
| n-Dodecane | $C_{12}H_{26}$ | –9.5°C | 216.2 | 748 |
| n-Tridecane | $C_{13}H_{28}$ | –5.3°C | 154.5 | 756 |

decreases as concentration increases. The mixtures of paraffin (dodecane and tridecane) have been applied to refrigeration. The melting temperature ranges from −17°C to −12°C, and the latent heat ranges from 165 to 185 kJ/kg without suffering subcooling [39]. In general, fewer organic homogenous PCMs are commercially available than inorganic homogenous PCMs.

### 3.2.2.3 Nanoadditive and Composite PCM

This type of PCMs is formed by adding complementary materials into the formula, such as nanoparticles, thickeners, or embedding the PCM in a matrix of porous material. Adding complementary materials can reduce subcooling, decrease nucleation time, increase thermal conductivity, and prevent leakage. On the contrary, these nanoparticles and matrixes reduce the latent heat, increase cost, and deteriorate stability. The pros and cons of adding complementary materials are listed in Table 3.10.

### 3.2.2.4 Macroencapsulated PCM

Macroencapsulation is to confine PCMs in containments from several milliliters to several liters. Usually, the diameter of the PCM core is above 1 mm. Thus far, it is the most widely used PCM encapsulation technique. Encapsulation can protect the PCMs from contamination and shape change, help enhance the heat transfer by a larger heat transfer area and reduce thermal resistance due to the limited thickness in a low-cost way [41,42]. The containment shape can be spherical, rectangular, cylindrical, pouches, flat plates, and other specially made shapes to fit the users' demands. The materials for making macroencapsulation can be metallic or plastic. Containments usually include an air cushion to compensate for the material volume change, especially for rigid capsules. The corrosion issue should be carefully chosen as the containment contacts with PCMs or/and HTFs directly (Figure 3.8).

### 3.2.2.5 Microencapsulated PCM

Microencapsulation refers to containing the PCMs in capsules at a micro-scale (few millimeters or less) to avoid changing the composition of the material due to contact with the environment. The encapsulation of PCM particles ranges 1–1,000 µm diameter with a continuous film shell. Capsule shells are usually made of polymers, such as polyamides, polyurethanes, polyureas, and polyesters. Since the PCMs are divided

---

**TABLE 3.10**

**Pros and Cons of Complementary Materials**

| Items | Pros | Cons |
| --- | --- | --- |
| Nanoparticles | Increase thermal conductivity, improve nucleation, reduce subcooling effect | Decrease latent heat, agglomerate over time, affect the uniformity of the material |
| Thickening agents | Prevent nucleating agent precipitation, reduce phase separation | Compromise thermal properties |
| Matrix composite | Improve thermal conductivity with low cost, prevent phase separation and leakage problem | Maximum filling ratio seems uneconomical as the gain in performance at a relatively high filling becomes insignificant [40] |

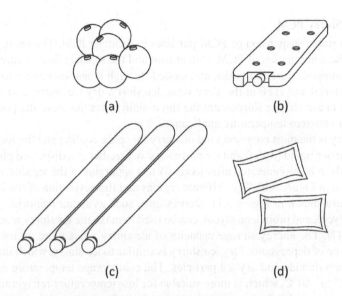

**FIGURE 3.8**  Common geometries for macroencapsulation of PCM: (a) spherical, (b) rectangular, (c) cylindrical, and (d) pouches [36].

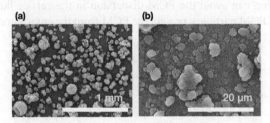

**FIGURE 3.9**  SEM micrographs of (a) the overall morphology of the PEMA microcapsules and (b) the surface roughness of the PEMA microcapsules [46].

into very small parts, the surface/volume is enhanced significantly, which improves heat transfer efficiency. Other advantages, such as mechanical stability, diminished agglomeration, minimized phase separation, and improved cycling stability, have been observed [36]. However, there is limited literature reporting the long-term thermal performance of microencapsulated PCM. The subcooling effect has been exacerbated by this technology, which is the major drawback of its application [43]. In addition, the destruction of the capsules during thermal cycles is one of the operation issues [44]. The manufacturing processes, no matter using the physical or chemical method, are still too complicated, which makes it suffer higher production costs [45].

Figure 3.9 illustrates the SEM micrographs of poly (ethylene-alt-maleic-anhydride) (PEMA) microcapsules. PEMA has long been used as a reactive foothold for the urea–formaldehyde polycondensation reaction. These images convey microcapsules with rough and porous exterior structures. These results indicate that low cross-linking density was exhibited for the shell, alluding to a porous structure. The observed payload of the PEMA microcapsules using heptane as PCM core is only 14.13% [46]. Therefore, the PEMA is more suitable for less volatile core materials. For volatile organic compounds, such as heptane, the produced microcapsules are more porous with poor barrier properties.

### 3.2.2.6  Slurry PCM

Slurry is a stable suspension of PCM particles in a carrier fluid. The energy can be stored in the combination of PCM's latent heat and the sensible heat of carrier fluid. It can be pumped, stored in tanks, and passed through a heat exchanger to work as storage material and HTF at the same time. Ice slurry, dry ice slurry, and microencapsulated phase change slurries are the three main types that have the potential to be used for sub-zero temperature applications.

Ice slurry is the most common kind of slurry for space cooling and the food industry. The carrier liquid (water) and ice particles stay together in a dispersed phase. The slurry needs to be re-generated after usage. A test apparatus of the ice slurry generator is shown in Figure 3.10. The different regions and time evolution of the ice slurry generator are shown in Figure 3.11. Depressants, such as sodium chloride, ethanol, ethylene glycol, and propylene glycol, can be used to make the ice slurry reach −30°C or lower [47]. The energy storage capacity of ice slurry relies more on ice fraction than the type of depressants. Dry ice slurry is similar to ice slurry, which consists of liquid carbon dioxide and dry ice particles. The cold storage temperature can range from −56°C to −80°C, which is more suitable for low-temperature refrigeration.

A microencapsulated slurry is comprised of liquid carriers and microencapsulated PCMs. The PCMs are packed inside the microencapsulation made by polymers. Microencapsulation can avoid the PCM dispersion in the carrier liquid, prevent the agglomeration of PCM particles, protect the PCM from the environment, and improve

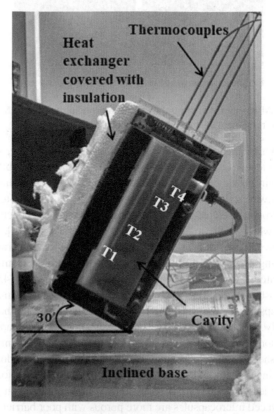

**FIGURE 3.10**   Test apparatus of ice slurry generator [48].

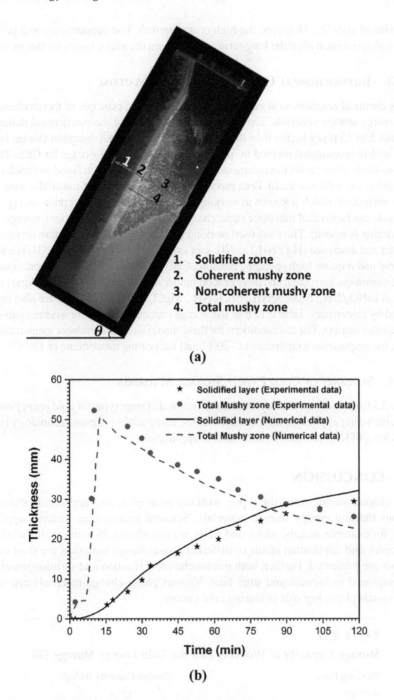

1. Solidified zone
2. Coherent mushy zone
3. Non-coherent mushy zone
4. Total mushy zone

(a)

(b)

**FIGURE 3.11** Result of ice slurry generator: (a) different regions formed during ice slurry generation, (b) evolution of solidified and mushy layer thickness [48].

mechanical stability. However, the high cost, relatively low conductivity, and performance deterioration after the long-term operation are the major issues for this method.

### 3.2.3 THERMOCHEMICAL COLD ENERGY STORAGE MATERIAL

Using chemical reactions to store and release energy is the principle of thermochemical cold energy storage materials. The energy storage capacity of thermochemical materials is about 5 to 15 times higher than that of sensible materials [49]. Sorption storage is the most widely investigated method to implement thermochemical storage for CES. There are two kinds of processes to implement this purpose: absorption (on liquid materials) and adsorption (on solid materials). Both processes involve two kinds of materials- a sorbent and a refrigerant, which is known as working pair. The conventional sorption energy storage cycle can be divided into three steps: charging (regeneration/desorption), storage, and discharging (sorption). The most used working pair for sub-zero temperature application is water and ammonia ($H_2O/NH_3$) as $NH_3$ can vapor below 0°C. However, $NH_3$ is a toxic material and requires high operating pressure. $CaCl_2/NH_3$ has a high adsorption capacity, while expansion, corrosion, and deterioration hinder its applications. Other working pairs, such as $LiNO_3/NH_3$, $NaSCN/NH_3$, $SrCl_2/NH_3$, $MgCl_2/NH3$, $PX21/NH_3$, are also investigated by researchers. Table 3.11 lists the storage capacity of different working pairs for cold energy storage. The test conditions for these materials are the ambient temperature of 35°C, the evaporation temperature of –20°C, and the heating temperature of 120°C.

### 3.2.4 SUMMARY OF COLD ENERGY STORAGE MATERIALS

Table 3.12 summarizes the key characteristics of different types of cold energy materials, including applicable temperature ranges, compatible systems, technology readiness level (TRL), pros, cons, and suitable applications.

## 3.3 CONCLUSION

This chapter summarized the types, working principles, and application status of various thermal energy storage materials. Sensible heat storage material applications, for example, metals, rocks, and water are introduced. In addition, the working principles and application status of different phase change materials are their compounds are presented. Further, both thermochemical reaction and sorption materials are employed to harness and store heat. Various phase-change materials and their compounds play a key role in storing cold energy.

**TABLE 3.11**
**Storage Capacity of Working Pair for Cold Energy Storage [50]**

| Working Pairs | Storage Capacity (kJ/kg) |
|---|---|
| $H_2O/NH_3$ | 122.4 |
| $LiNO_3/NH_3$ | 129.6 |
| $CaCl_2/NH_3$ | 295.2 |
| $SrCl_2/NH_3$ | 432.0 |
| $NaSCN/NH_3$ | 45.36 |
| $PX21/NH_3$ | 97.6 |

**TABLE 3.12**
**Summary of Cold Energy Storage Materials [36]**

| Material Type | Temperature Range Mentioned in the Literature | Compatible CES Types | Current Development Stage and Technology Readiness Level (TRL) | Pros | Cons/Technical Challenges | Suitable Applications |
|---|---|---|---|---|---|---|
| Sensible materials (solids) | −170°C to 0°C | Packed-bed and thermocline | Commercially available for some applications TRL: 8–9 | Low cost and high reliability, materials widely available, wide temperature range, suitable for large scale storage, environmentally friendly | Lower energy and exergy density, larger pressure drop | Large-scale electricity storage, large-scale refrigeration, large-scale waste cold recovery |
| Sensible materials (liquids) | −153°C to 0°C | Packed-bed and thermocline | Some candidate materials were selected but have not been widely applied TRL: 2–5 | Can act as the HTF and the thermal energy storage material at the same time | Working temperature range is limited by freezing and boiling point, low energy, and exergy density, can be flammable, corrosive, and toxic, become viscous at cryogenic temperatures | Large-scale electricity storage, large-scale refrigeration, large-scale waste cold recovery |
| Pure substances PCMs | −210°C to 0°C | Packed-bed and thermocline, shell-and-tube, plate-shaped | Some materials are already deployed commercially TRL: 5–9 | High energy and exergy density, wide temperature range | Can be flammable, corrosive, and toxic, need to be stored in pressurized or open containers for solid–gas/liquid–gas phase transition, need to handle subcooling and volume change | Small- and large-scale active and passive refrigeration, large-scale waste cold recovery, small- and large-scale electricity storage |

*(Continued)*

**TABLE 3.12 (*Continued*)**
**Summary of Cold Energy Storage Materials [36]**

| Material Type | Temperature Range Mentioned in the Literature | Compatible CES Types | Current Development Stage and Technology Readiness Level (TRL) | Pros | Cons/Technical Challenges | Suitable Applications |
|---|---|---|---|---|---|---|
| Homogeneous mixture PCMs | −86°C to 0°C | Packed-bed and thermocline, shell-and-tube, plate-shaped | Many eutectic solutions are commercially available, more under development TRL: 5–9 | High energy and exergy density; tunable freezing point | Need to handle subcooling, volume change, and phase separation due to incongruent melting | Small-scale active and passive refrigeration, small-scale waste cold recovery, small-and large-scale refrigeration |
| Nanoadditive and composite PCMs | −114°C to 0°C | Packed-bed and thermocline, shell-and-tube, plate-shaped | Additives for salt mixture and paraffin available, more under development TRL: 3–9 | Improves the thermal conductivity, avoids phase separation, reduces of subcooling effect, increases the growth rate | Reduces latent heat, increases cost, particles tend to agglomerate over time | Small- and large-scale active and passive refrigeration, small- and large-scale waste cold recovery |
| Microencapsulation and other slurry PCMs | −140°C to 0°C | Packed-bed and thermocline, shell-and-tube, plate-shaped, slurry-based | Ice slurry for refrigeration is commercially available, dry ice slurry is a conceptual phase, more microencapsulation is under development TRL: 3–6 | Better heat capacity and heat transfer coefficient, avoids agglomeration, slurries can be used as the storage and HTF at the same time, able to be pumped and have the combined energy density of the PCM latent heat and sensible heat of the carrier fluid | Microcapsulation related: high cost, microcapsulation can be destroyed at mid/long term, increase subcooling. Slurry related: stratification issue, heavy mechanical power required to generate small and | Small- and large-scale active refrigeration, small- and large-scale waste cold recovery |

*(Continued)*

**TABLE 3.12 (Continued)**
**Summary of Cold Energy Storage Materials [36]**

| Material Type | Temperature Range Mentioned in the Literature | Compatible CES Types | Current Development Stage and Technology Readiness Level (TRL) | Pros | Cons/Technical Challenges | Suitable Applications |
|---|---|---|---|---|---|---|
| | | | | | smooth ice crystals and to maintain the ice slurry in the homogeneous state, lack of full understanding of the basic physical properties | |
| Thermochemical materials | −30°C to 0°C | Absorption and adsorption cycles | Ammonia absorption is commercially available, other types in the early stages TRL: 1–9 | Higher energy and exergy density | High system complexity, toxic and corrosive, may require high pressure to operate, low heat and mass transfers in adsorbent beds | Small- and large-scale active refrigeration |

## REFERENCES

[1] Ayaz H, Chinnasamy V, Yong J, Cho H. Review of technologies and recent advances in low-temperature sorption thermal storage systems. *Energies* 2021;14. https://doi.org/10.3390/en14196052

[2] Li B, Li J, Shao H, He L. Mg-based hydrogen absorbing materials for thermal energy storage—A review. *Applied Sciences* 2018;8:1375. https://doi.org/10.3390/app8081375

[3] Cruickshank CA, Baldwin C. 19-Sensible thermal energy storage: Diurnal and seasonal. In: Letcher TM, editor. *Storing energy*, Amsterdam: Elsevier; 2022, pp. 419–41. https://doi.org/10.1016/B978-0-12-824510-1.00018-0

[4] Nash AL, Badithela A, Jain N. Dynamic modeling of a sensible thermal energy storage tank with an immersed coil heat exchanger under three operation modes. *Applied Energy* 2017; 195:877–89. https://doi.org/10.1016/j.apenergy.2017.03.092

[5] Al Edhari A, Ngo C. Experimental study of thermal energy storage using natural porous media. *Proceedings of the ASME 2017 International Mechanical Engineering Congress and Exposition. Volume 8: Heat Transfer and Thermal Engineering.* Tampa, Florida, USA. November 3–9, 2017. V008T10A052. ASME. https://doi.org/10.1115/IMECE2017-71714

[6] Voronin D, Ivanov E, Gushchin P, Fakhrullin R, Vinokurov V. Clay composites for thermal energy storage: A review. *Molecules* 2020;25(7):1504. https://doi.org/10.3390/molecules25071504

[7] Sarier N, Onder E. Organic phase change materials and their textile applications: An overview. *Thermochimica Acta* 2012;540:7–60. https://doi.org/10.1016/j.tca.2012.04.013

[8] Kenisarin MM. Thermophysical properties of some organic phase change materials for latent heat storage: A review. *Solar Energy* 2014;107:553–75. https://doi.org/10.1016/j.solener.2014.05.001

[9] Himran S, Suwono A, Mansoori GA. Characterization of alkanes and paraffin waxes for application as phase change energy storage medium. *Energy Sources* 1994;16:117–28. https://doi.org/10.1080/00908319408909065

[10] Dirand M, Bouroukba M, Briard A-J, Chevallier V, Petitjean D, Corriou J-P. Temperatures and enthalpies of (solid + solid) and (solid + liquid) transitions of n-alkanes. *The Journal of Chemical Thermodynamics* 2002;34:1255–77. https://doi.org/10.1006/jcht.2002.0978

[11] Rozanna D, Chuah TG, Salmiah A, Choong TSY, Sa'ari M. Fatty acids as phase change materials (PCMs) for thermal energy storage: A review. *International Journal of Green Energy* 2005;1:495–513. https://doi.org/10.1081/GE-200038722

[12] Sharma SD, Sagara K. Latent heat storage materials and systems: A review. *International Journal of Green Energy* 2005;2:1–56. https://doi.org/10.1081/GE-200051299

[13] Gunasekara SN, Pan R, Chiu JN, Martin V. Polyols as phase change materials for surplus thermal energy storage. *Applied Energy* 2016;162:1439–52. https://doi.org/10.1016/j.apenergy.2015.03.064

[14] Junaid MF, Rehman Z ur, Čekon M, Čurpek J, Farooq R, Cui H. Inorganic phase change materials in thermal energy storage: A review on perspectives and technological advances in building applications. *Energy and Buildings* 2021;252:111443. https://doi.org/10.1016/j.enbuild.2021.111443

[15] Singh P, Sharma RK, Ansu AK, Goyal R, Sarı A, Tyagi VV. A comprehensive review on development of eutectic organic phase change materials and their composites for low and medium range thermal energy storage applications. *Solar Energy Materials and Solar Cells* 2021;223:110955. https://doi.org/10.1016/j.solmat.2020.110955

[16] Fauzi H, Metselaar HSC, Mahlia TMI, Silakhori M, Nur H. Phase change material: Optimizing the thermal properties and thermal conductivity of myristic acid/palmitic acid eutectic mixture with acid-based surfactants. *Applied Thermal Engineering* 2013;60:261–5. https://doi.org/10.1016/j.applthermaleng.2013.06.050

[17] Kant K, Shukla A, Sharma A. Ternary mixture of fatty acids as phase change materials

for thermal energy storage applications. *Energy Reports* 2016;2:274–9. https://doi.org/10.1016/j.egyr.2016.10.002

[18] Yu W, France DM, Routbort JL, Choi SUS. Review and comparison of nanofluid thermal conductivity and heat transfer enhancements. *Heat Transfer Engineering* 2008;29:432–60. https://doi.org/10.1080/01457630701850851

[19] Khadiran T, Hussein MZ, Zainal Z, Rusli R. Encapsulation techniques for organic phase change materials as thermal energy storage medium: A review. *Solar Energy Materials and Solar Cells* 2015;143:78–98. https://doi.org/10.1016/j.solmat.2015.06.039

[20] Wang C, Feng L, Li W, Zheng J, Tian W, Li X. Shape-stabilized phase change materials based on polyethylene glycol/porous carbon composite: The influence of the pore structure of the carbon materials. *Solar Energy Materials and Solar Cells* 2012;105:21–6. https://doi.org/10.1016/j.solmat.2012.05.031

[21] Hu H. Recent advances of polymeric phase change composites for flexible electronics and thermal energy storage system. *Composites Part B: Engineering* 2020;195:108094. https://doi.org/10.1016/j.compositesb.2020.108094

[22] Wang Q, Zhou D, Chen Y, Eames P, Wu Z. Characterization and effects of thermal cycling on the properties of paraffin/expanded graphite composites. *Renewable Energy* 2020;147:1131–8. https://doi.org/10.1016/j.renene.2019.09.091

[23] Zhang Z, Fang X. Study on paraffin/expanded graphite composite phase change thermal energy storage material. *Energy Conversion and Management* 2006;47:303–10. https://doi.org/10.1016/j.enconman.2005.03.004

[24] Chakraborty A, Noh J, Mach R, Shamberger P, Yu C. Thermal energy storage composites with preformed expanded graphite matrix and paraffin wax for long-term cycling stability and tailored thermal properties. *Journal of Energy Storage* 2022;52:104856. https://doi.org/10.1016/j.est.2022.104856

[25] Ling Z, Chen J, Xu T, Fang X, Gao X, Zhang Z. Thermal conductivity of an organic phase change material/expanded graphite composite across the phase change temperature range and a novel thermal conductivity model. *Energy Conversion and Management* 2015;102:202–8. https://doi.org/10.1016/j.enconman.2014.11.040

[26] Aydin D, Casey SP, Riffat S. The latest advancements on thermochemical heat storage systems. *Renewable and Sustainable Energy Reviews* 2015;41:356–67. https://doi.org/10.1016/j.rser.2014.08.054

[27] Pardo P, Deydier A, Anxionnaz-Minvielle Z, Rougé S, Cabassud M, Cognet P. A review on high temperature thermochemical heat energy storage. *Renewable and Sustainable Energy Reviews* 2014;32:591–610. https://doi.org/10.1016/j.rser.2013.12.014

[28] Jarimi H, Aydin D, Yanan Z, Ozankaya G, Chen X, Riffat S. Review on the recent progress of thermochemical materials and processes for solar thermal energy storage and industrial waste heat recovery. *International Journal of Low-Carbon Technologies* 2019;14:44–69. https://doi.org/10.1093/ijlct/cty052

[29] Henninger SK, Jeremias F, Kummer H, Schossig P, Henning H-M. Novel sorption materials for solar heating and cooling. *Energy Procedia* 2012;30:279–88. https://doi.org/10.1016/j.egypro.2012.11.033

[30] Srivastava NC, Eames IW. A review of adsorbents and adsorbates in solid–vapour adsorption heat pump systems. *Applied Thermal Engineering* 1998;18:707–14. https://doi.org/10.1016/S1359-4311(97)00106-3

[31] Sarbu I, Sebarchievici C. A comprehensive review of thermal energy storage. *Sustainability* 2018;10. https://doi.org/10.3390/su10010191

[32] Davenne TR, Garvey SD, Cardenas B, Simpson MC. The cold store for a pumped thermal energy storage system. *Journal of Energy Storage* 2017;14:295–310. https://doi.org/10.1016/j.est.2017.03.009

[33] Sharma A, Tyagi VV, Chen CR, Buddhi D. Review on thermal energy storage with phase change materials and applications. *Renewable and Sustainable Energy Reviews* 2009;13:318–45. https://doi.org/10.1016/j.rser.2007.10.005

[34] Li G, Hwang Y, Radermacher R, Chun H-H. Review of cold storage materials for sub-zero applications. *Energy* 2013;51:1–17. https://doi.org/10.1016/j.energy.2012.12.002

[35] Oró E, de Gracia A, Castell A, Farid MM, Cabeza LF. Review on phase change materials (PCMs) for cold thermal energy storage applications. *Applied Energy* 2012;99:513–33. https://doi.org/10.1016/j.apenergy.2012.03.058

[36] Yang L, Villalobos U, Akhmetov B, Gil A, Khor JO, Palacios A, et al. A comprehensive review on sub-zero temperature cold thermal energy storage materials, technologies, and applications: State of the art and recent developments. *Applied Energy* 2021;288:116555. https://doi.org/10.1016/j.apenergy.2021.116555

[37] Rasta IM, Suamir IN. Study on thermal properties of bio-PCM candidates in comparison with propylene glycol and salt based PCM for sub-zero energy storage applications. *IOP Conference Series: Materials Science and Engineering* 2019;494:012024. https://doi.org/10.1088/1757-899x/494/1/012024

[38] Sze JY, Mu C, Romagnoli A, Li Y. Non-eutectic phase change materials for cold thermal energy storage. *Energy Procedia* 2017;143:656–61. https://doi.org/10.1016/j.egypro.2017.12.742

[39] Gunasekara SN, Kumova S, Chiu JN, Martin V. Experimental phase diagram of the dodecane–tridecane system as phase change material in cold storage. *International Journal of Refrigeration* 2017;82:130–40. https://doi.org/10.1016/j.ijrefrig.2017.06.003

[40] Zhu Z-Q, Huang Y-K, Hu N, Zeng Y, Fan L-W. Transient performance of a PCM-based heat sink with a partially filled metal foam: Effects of the filling height ratio. *Applied Thermal Engineering* 2018;128:966–72. https://doi.org/10.1016/j.applthermaleng.2017.09.047

[41] Aftab W, Huang X, Zou R. The application of carbon materials in latent heat thermal energy storage (LHTES). In: Zhang G, editor. *Thermal transport in carbon-based nanomaterials*, Amsterdam: Elsevier; 2017, pp. 243–65. https://doi.org/10.1016/B978-0-32-346240-2.00009-1

[42] Höhlein S, König-Haagen A, Brüggemann D. Macro-encapsulation of inorganic phase-change materials (PCM) in metal capsules. *Materials* 2018;11(9), 1752. https://doi.org/10.3390/ma11091752

[43] Fan YF, Zhang XX, Wang XC, Li J, Zhu QB. Super-cooling prevention of microencapsulated phase change material. *Thermochimica Acta* 2004;413:1–6. https://doi.org/10.1016/j.tca.2003.11.006

[44] Youssef Z, Delahaye A, Huang L, Trinquet F, Fournaison L, Pollerberg C. State of the art on phase change material slurries. *Energy Conversion and Management* 2013;65:120–32. https://doi.org/10.1016/j.enconman.2012.07.004

[45] Zhang P, Ma ZW. An overview of fundamental studies and applications of phase change material slurries to secondary loop refrigeration and air conditioning systems. *Renewable and Sustainable Energy Reviews* 2012; 16: 5021–58. https://doi.org/10.1016/j.rser.2012.03.059

[46] Mustapha AN, Zhang Y, Zhang Z, Ding Y, Li Y. A systematic study on the reaction mechanisms for the microencapsulation of a volatile phase change material (PCM) via one-step in situ polymerisation. *Chemical Engineering Science* 2022;252:117497. https://doi.org/10.1016/j.ces.2022.117497

[47] Kauffeld M, Wang MJ, Goldstein V, Kasza KE. Ice slurry applications. *International Journal of Refrigeration* 2010;33:1491–505. https://doi.org/10.1016/j.ijrefrig.2010.07.018

[48] Tiwari VK, Kumar A, Kumar A. Enhancing ice slurry generation by using inclined cavity for subzero cold thermal energy storage: Simulation, experiment and performance analysis. *Energy* 2019;183:398–414. https://doi.org/10.1016/j.energy.2019.06.121

[49] Dincer I, Ezan MA. *Heat storage: A unique solution for energy systems*. Cham: Springer; 2018.

[50] Mugnier D, Goetz V. Energy storage comparison of sorption systems for cooling and refrigeration. *Solar Energy* 2001;71:47–55. https://doi.org/10.1016/S0038-092X(01)00013-5

# 4 Integrated Waste Thermal Energy Recovery Systems

## 4.1 SYSTEM APPLICATIONS ON WASTE HEAT RECOVERY

### 4.1.1 INTEGRATED SYSTEMS WITH SINGLE PRODUCT

#### 4.1.1.1 A Cascade Absorption Refrigeration System

Temperature/energy cascade systems are designed to improve waste heat recovery efficiency. Cui et al. [1] proposed a cascade $NH_3$–$H_2O$/$LiBr$–$H_2O$ absorption refrigeration system to harness waste heat spanning 90–150°C. The waste heat at higher temperatures is used to operate the $NH_3$–$H_2O$ absorption refrigeration system. Comparatively, the waste heat at a lower temperature is employed to operate the $LiBr$–$H_2O$ absorption refrigeration system. The cooling effect produced by the $LiBr$–$H_2O$ absorption refrigeration system removes the heat generated in the $NH_3$–$H_2O$ absorption refrigeration system.

Figure 4.1 depicts the schematic diagram of the cascade absorption refrigeration system. The proposed cascade absorption refrigeration system comprises an $NH_3$–$H_2O$ absorption and a $LiBr$–$H_2O$ absorption system. First, the $NH_3$ is heated in the generator using the available waste heat. The heated $NH_3$ gas condenses into a liquid $NH_3$–$H_2O$ solution in the condenser. Then, the $NH_3$–$H_2O$ solution evaporates in the evaporator, resulting in a cooling effect. Consequently, the $NH_3$–$H_2O$ solution is diluted and flows into the $NH_3$ storage. In the absorber, the $NH_3$–$H_2O$ solution absorbs $NH_3$ that evaporates from the evaporator. Eventually, the diluted $NH_3$–$H_2O$ solution flows into the generator. During the entire process, the cooling effect produced by the $LiBr$–$H_2O$ absorption system is employed to remove the heat from the condenser and absorber of the $NH_3$–$H_2O$ absorption system. The working principles behind the $LiBr$–$H_2O$ absorption system have been introduced in Chapter 2. The designed thermal properties of the cascade absorption system are listed in Table 4.1.

The energy and mass balance equations in the LiBr generator are expressed as:

$$m_1 = m_2 + m_3 \tag{4.1}$$

$$m_1 x_1 = m_3 x_3 \tag{4.2}$$

$$Q_{gen,LiBr} = m_2 h_2 + m_3 h_3 - m_1 h_1 \tag{4.3}$$

The energy and mass balance equations in the LiBr absorber are presented as:

$$m_9 = m_5 + m_8 \tag{4.4}$$

DOI: 10.1201/9781003343387-4

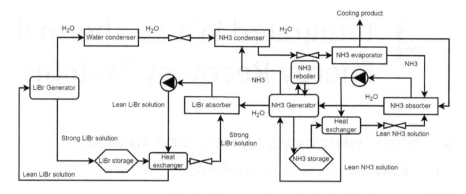

**FIGURE 4.1**   Schematic diagram of a LiBr/NH$_3$ cascade absorption refrigeration system [1].

**TABLE 4.1**
**Thermal Properties of the Cascade NH$_3$–H$_2$O/LiBr–H$_2$O Absorption Chiller [1]**

|  | Temperature (°C) | Pressure (kPa) | Mass Flow (kg/h) | Mass Fraction | | |
|---|---|---|---|---|---|---|
|  |  |  |  | H$_2$O | Li$^+$ | Br |
| 1 | 66.1 | 9.65 | 53,750 | 0.5 | 0.04 | 0.46 |
| 2 | 80.0 | 9.65 | 4,903.9 | 1 | 0 | 0 |
| 3 | 80.0 | 9.65 | 4,8846.1 | 0.45 | 0.044 | 0.506 |
| 4 | 39.0 | 9.65 | 4,8846.1 | 0.45 | 0.044 | 0.506 |
| 5 | 39.0 | 1.292 | 4,8846.1 | 0.45 | 0.044 | 0.506 |
| 6 | 45.1 | 9.65 | 4,903.9 | 1 | 0 | 0 |
| 7 | 10.0 | 1.292 | 4,903.9 | 1 | 0 | 0 |
| 8 | 10.0 | 1.292 | 4,903.9 | 1 | 0 | 0 |
| 9 | 34.0 | 9.65 | 53,750 | 0.5 | 0.04 | 0.46 |
| 10 | 34.0 | 9.65 | 53,750 | 0.5 | 0.04 | 0.46 |
|  |  |  |  | H$_2$O | NH$_3$ |  |
| 11 | 72.9 | 724.4 | 15,063 | 0.659 | 0.341 |  |
| 12 | 67.6 | 724.4 | 3,205 | trace | 1 |  |
| 13 | 118.0 | 724.4 | 11,858 | 0.837 | 0.163 |  |
| 14 | 42.0 | 724.4 | 11,858 | 0.837 | 0.163 |  |
| 15 | 42.0 | 72.2 | 11,858 | 0.837 | 0.163 |  |
| 16 | 15.0 | 724.4 | 3,205 | trace | 1 |  |
| 17 | −40.0 | 72.2 | 3,205 | trace | 1 |  |
| 18 | −40.0 | 72.2 | 3,205 | trace | 1 |  |
| 19 | 15.0 | 72.2 | 15,063 | 0.659 | 0.341 |  |
| 20 | 15.0 | 724.4 | 15,063 | 0.659 | 0.341 |  |

$$m_9 x_9 = m_5 x_5 \tag{4.5}$$

$$Q_{abs,LiBr} = m_9 h_9 + m_5 h_5 - m_8 h_8 \tag{4.6}$$

The energy and mass balance equations in the LiBr condenser are expressed as:

$$m_6 = m_2 \tag{4.7}$$

$$Q_{con,LiBr} = m_6 h_6 - m_2 h_2 \tag{4.8}$$

The energy and mass balance equations in the LiBr evaporator are presented as:

$$m_8 = m_7 \tag{4.9}$$

$$Q_{eva,LiBr} = m_8 h_8 - m_7 h_7 \tag{4.10}$$

The energy and mass balance equations in the LiBr heat exchanger are expressed as:

$$m_4 + m_1 = m_3 + m_{10} \tag{4.11}$$

$$m_4 h_4 - m_3 h_3 = m_1 h_1 - m_{10} h_{10} \tag{4.12}$$

The energy and mass balance equations in the $NH_3$ generator are given as:

$$m_{11} = m_{12} + m_{13} \tag{4.13}$$

$$m_{11} x_{11} = m_{12} x_{12} + m_{13} x_{13} \tag{4.14}$$

$$Q_{gen,NH3} = m_{12} h_{12} + m_{13} h_{13} - m_{11} h_{11} \tag{4.15}$$

The energy and mass balance equations in the $NH_3$ absorber are presented as:

$$m_{19} = m_{15} + m_{18} \tag{4.16}$$

$$m_{19} x_{19} = m_{15} x_{15} + m_{18} x_{18} \tag{4.17}$$

$$Q_{abs,NH3} = m_{19} h_{19} + m_{15} h_{15} - m_{18} h_{18} \tag{4.18}$$

The energy and mass balance equations in the $NH_3$ condenser are expressed as:

$$m_{16} = m_{12} \tag{4.19}$$

$$Q_{con,NH3} = m_{16} h_{16} - m_{12} h_{12} \tag{4.20}$$

The energy and mass balance equations in the $NH_3$ evaporator are given as:

$$m_{18} = m_{17} \tag{4.21}$$

$$Q_{eva,NH_3} = m_{18}h_{18} - m_{17}h_{17} \tag{4.22}$$

The energy and mass balance equations in the LiBr heat exchanger are presented as:

$$m_{14} + m_{11} = m_{13} + m_{20} \tag{4.23}$$

$$m_{14}h_{14} - m_{13}h_{13} = m_{11}h_{11} - m_{20}h_{20} \tag{4.24}$$

### 4.1.1.2   A $CO_2$–$NH_3$ Cascade Refrigeration System

Alberto Dopazo and coworkers [2] proposed a $CO_2$–$NH_3$ cascade refrigeration system. A cascade heat exchanger is employed to connect the $NH_3$ and $CO_2$ circuits. The waste heat at a higher temperature is used to thermally drive the $NH_3$ circuit. The waste heat at a lower temperature is employed to operate the $CO_2$ circuit. The cooling effect produced by the $NH_3$ circuit is employed to cool the $CO_2$ in the condenser.

The cooling effect of the cascade refrigeration system is produced by the evaporator of the $CO_2$ circuit. The gas $CO_2$ is compressed and released into the condenser. Consequently, the gas $CO_2$ condenses in the condenser as the evaporator in the $NH_3$ circuit lowers the condenser temperature. Thereafter, the condensed $CO_2$ flows into the evaporator to produce the cooling effect. The working principles of the $NH_3$ circuit are similar to the $CO_2$ circuit. A cooling stream of water is required to remove heat from the condenser in the $NH_3$ circuit.

The energy and mass balance equations in the $NH_3$-compressor are expressed as:

$$m_6 = m_5 \tag{4.25}$$

$$W_{comp,NH_3} = \frac{m_1(h_{6s} - h_5)}{\eta_{comp,NH_3}} \tag{4.26}$$

The energy and mass balance equations in the $NH_3$ expansion device are presented as:

$$m_8 = m_7 \tag{4.27}$$

$$h_8 = h_7 \tag{4.28}$$

The energy and mass balance equations in the $NH_3$ condenser are expressed as:

$$m_6 = m_7 \tag{4.29}$$

$$Q_{con,NH_3} = m_5(h_7 - h_6) \tag{4.30}$$

The energy and mass balance equations in the $CO_2$ compressor are given as:

$$m_2 = m_1 \tag{4.31}$$

$$W_{comp,co_2} = \frac{m_1(h_{2s} - h_1)}{\eta_{comp,co_2}} \tag{4.32}$$

The energy and mass balance equations in the $CO_2$ expansion device are expressed as:

$$m_3 = m_4 \tag{4.33}$$

$$h_3 = h_4 \tag{4.34}$$

The energy and mass balance equations in the $CO_2$ evaporator are presented as:

$$m_1 = m_4 \tag{4.35}$$

$$Q_{eva,co_2} = m_1(h_1 - h_4) \tag{4.36}$$

The energy and mass balance equations in the cascade heat exchanger are expressed as:

$$m_2 = m_3 \tag{4.37}$$

$$m_8 = m_5 \tag{4.38}$$

$$m_1(h_3 - h_2) = m_5(h_5 - h_8) \tag{4.39}$$

The COP of the $CO_2$–$NH_3$ cascade refrigeration system is defined as the ratio of the cooling effects produced by the $CO_2$ circuit with respect to the total electricity consumption and is expressed as (Figure 4.2):

$$COP = \frac{Q_{c,co_2}}{E_{com,co_2} + E_{com,NH_3}} \tag{4.40}$$

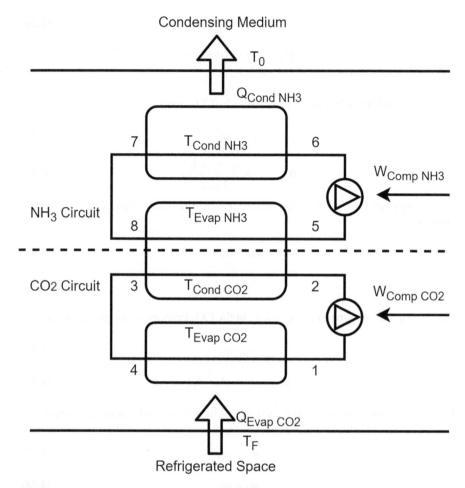

**FIGURE 4.2** (a) Schematic diagram of a $CO_2/NH_3$ cascade refrigeration system; log P–h diagram of the $CO_2$ and the $NH_3$ circuits [2].

### 4.1.1.3 A Cascade Adsorption Chiller System

Dakkama et al. [3] proposed a cascade adsorption refrigeration system to produce low-temperature cooling by utilizing waste heat. The performance of the system, which employed five pairs of working materials, is investigated. Figure 4.3 portrays a schematic diagram of the cascade adsorption chiller system. The cascade adsorption refrigeration system comprises four adsorbers, one condenser, one evaporator, and one integrated evaporator/condenser heat exchanger. High-temperature waste heat is employed to raise the temperature of the upper adsorber. The produced cooling is employed to cool the bottom adsorption refrigeration system; utilizing the waste heat at low temperatures as the heat source. Figure 4.3 illustrates the temperature profiles of the cascade adsorption chiller system. Results reveal that the lowest chilled water temperature is able to reach $-10°C$.

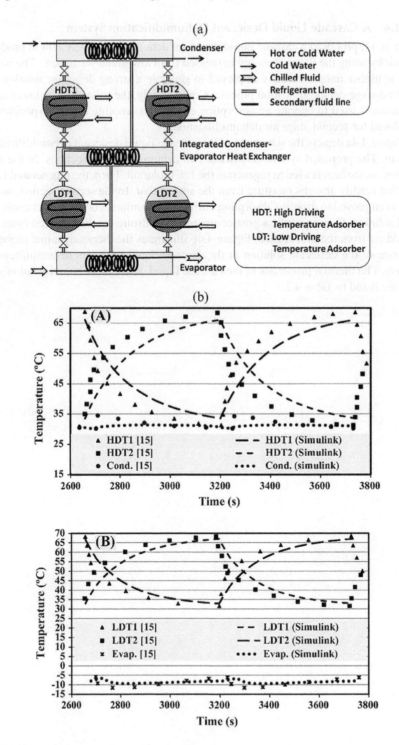

**FIGURE 4.3** (a) Schematic diagram of a cascade adsorption chiller system; (b) temperature profile of the cascade adsorption chiller system [3].

#### 4.1.1.4    A Cascade Liquid Desiccant Dehumidification System

Su et al. [4] proposed a cascade liquid desiccant dehumidification system to produce dry air by using the thermal energy harnessed from low-grade waste heat. The waste heat at higher temperatures is employed to generate a strong desiccant solution for the first-stage dehumidification process. Contrastingly, the waste heat at lower temperatures is used to operate an absorption chiller. The cooling effect is specifically employed for second-stage air dehumidification.

Figure 4.4a depicts the schematic of the cascade liquid desiccant dehumidification system. The proposed system consists of two dehumidification sections. In the first section, waste heat is used to regenerate the LiCl solution. Then, the regenerated LiCl solution readily absorbs moisture from the ambient air. In the second section, waste heat is employed to drive an absorption chiller. The resultant cooling effect cools the LiCl solution. Consequently, a greater amount of moisture can be removed from the humid air from the first section. Figure 4.4b illustrates the thermodynamic property diagram of the desiccant solution in the cascade liquid desiccant dehumidification system. The thermal properties of the cascade liquid desiccant dehumidification system are listed in Table 4.2.

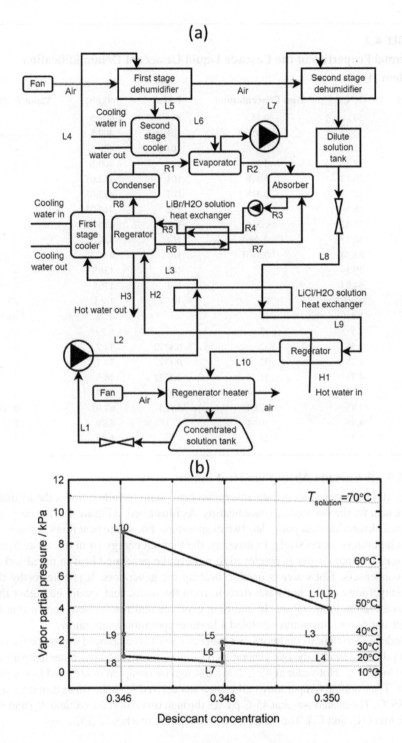

**FIGURE 4.4** (a) Schematic of a cascade liquid desiccant dehumidification system. (b) thermodynamic diagram of desiccant solution in the cascade liquid desiccant dehumidification system [4].

**TABLE 4.2**

**Thermal Properties of the Cascade Liquid Desiccant Dehumidification System [4]**

| Point | T(°C) | Mass Concentration | P(kPa) | m(kg/s) | Vapor Fraction |
|-------|-------|--------------------|--------|---------|----------------|
| L1 | 50 | 0.35 | 81 | 409.5 | – |
| L2 | 50 | 0.35 | 101 | 409.5 | – |
| L3 | 35.36 | 0.35 | 101 | 409.5 | – |
| L4 | 32 | 0.35 | 101 | 409.5 | – |
| L5 | 36.13 | 0.348 | 101 | 412.05 | – |
| L6 | 32 | 0.348 | 101 | 412.05 | – |
| L7 | 18 | 0.348 | 101 | 412.05 | – |
| L8 | 25.42 | 0.3461 | 101 | 414.32 | – |
| L9 | 40.17 | 0.3461 | 101 | 414.32 | – |
| L10 | 65.24 | 0.3461 | 101 | 414.32 | – |
| H1 | 70.24 | – | 101 | 1365 | – |
| H2 | 64.93 | – | 101 | 1365 | – |
| H3 | 61.68 | – | 101 | 1365 | – |
| R1 | 1.5 | 13 | – | 6.79 | 0.03 |
| R2 | 1.5 | 13 | – | 6.79 | 1 |
| R3 | 1.5 | 32 | 0.4772 | 74.7 | 0 |
| R4 | 4.75 | 32 | 0.4772 | 74.7 | 0 |
| R5 | 4.75 | 51.75 | 0.4772 | 74.7 | 0 |
| R6 | 4.75 | 59.94 | 0.525 | 67.91 | 0 |
| R7 | 4.75 | 37.03 | 0.525 | 67.91 | 0 |
| R8 | 4.75 | 59.94 | – | 6.79 | 1 |
| R9 | 4.75 | 32 | – | 6.79 | 1 |

### 4.1.1.5 A Cascade Absorption Heat Pump

Xu et al. [5] proposed a cascade absorption heat pump to fully harness the available waste heat for the purpose of space heating. As illustrated in Figure 4.5, the proposed cascade absorption heat pump has two evaporators. The waste heat flows through the two evaporators successively. In this way, the thermal energy from the waste heat is recovered. Heated water is employed to remove the generated heat in the absorbers and condensers. Hot water is used to heat up the generators. It is noteworthy that the temperature of the hot water derived from the waste heat should be higher than the evaporation temperature. In addition, experimental results have shown that the higher hot water temperature enabled a lower evaporation temperature.

Under experimental conditions, hot water at 120.1°C is employed to provide the necessary thermal heat to the two generators ($G_1$ and $G_2$). The outlet hot water temperature drops to 83.6°C. Hot water at 45°C flows through two evaporators ($E_1$ and $E_2$) successively. The eventual temperature of the hot water derived from the waste heat decreases to 15.9°C. The heated water at 45°C passes through two absorbers ($A_1$ and $A_2$) and two condensers ($C_1$ and $C_2$). The outlet water temperature reaches 52.5°C.

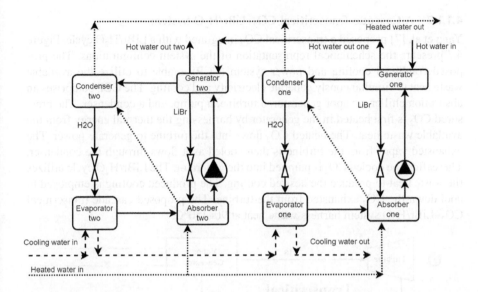

**FIGURE 4.5**   Schematical diagram of a cascade heat pump heating system [5].

### 4.1.1.6   Combined ICE, ORC, and KC Cycle

He et al. [6] proposed a combined organic Rankine cycle and Kalina cycle system to harness the thermal energy dissipated by an internal combustion engine. The working principles of the organic Rankine cycle and Kalina cycle have been presented in Chapter 2. The system configuration is introduced in Figure 4.6. In the organic Rankine cycle, the working fluid becomes a high-pressure vapor by harnessing the thermal energy from the super-heated exhausted gas. The generated vapor then flows into the turbine to generate power. The exhausted vapor is employed to heat up the working fluid in the Kalina cycle. The combined organic Rankine cycle and Kalina cycle system has demonstrated a better waste heat recovery efficiency when compared with a single cycle [6].

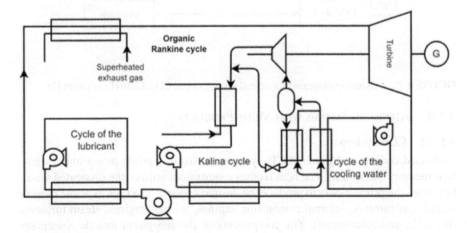

**FIGURE 4.6**   System diagram of the combined thermodynamic cycle for waste heat recovery of ICE.

#### 4.1.1.7 A Cascade Transcritical $CO_2$–LiBr/$H_2O$ System

Yang et al. [7] proposed a transcritical $CO_2$ integrated with a LiBr/$H_2O$ cycle. Figure 4.7 presents the schematical representation of the system configurations. The proposed combined cooling and power system (CCP) is able to utilize the available waste heat to simultaneously generate electricity and cooling. The CCP comprises an absorption chiller, a vapor generator, a turbine, a pump, and a condenser. The pressured $CO_2$ is first heated in the generator by harnessing the thermal energy from the available waste heat. The heated $CO_2$ flows into the turbine to generate power. The exhausted vapor from the turbine is then cooled as it flows through the condenser. Thereafter, the cooled $CO_2$ is pumped into the generator. The LiBr/$H_2O$ cycle utilizes the waste heat to produce the desired cooling. The produced cooling is employed to cool down the $CO_2$ exhausted from the turbine. The proposed cascade transcritical $CO_2$–LiBr/$H_2O$ system harness waste heat at 90–150°C.

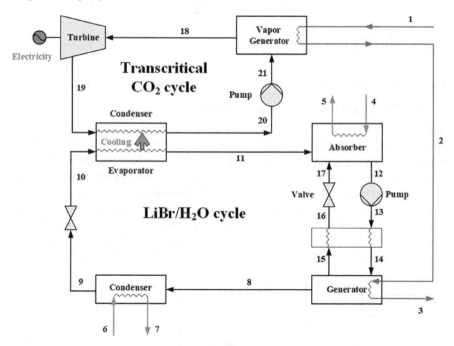

**FIGURE 4.7** Schematic diagram of a cascade transcritical $CO_2$–LiBr/$H_2O$ system [7].

### 4.1.2 INTEGRATED SYSTEMS WITH MULTI-PRODUCTS

#### 4.1.2.1 CCP Systems

Combined cooling and power (CCP) systems consist of a prime mover and a waste heat recovery subsystem. The heat recovery subsystem utilizes the dissipated waste heat from the prime mover to produce the desired cooling. The typical prime movers include gas turbines, internal combustion engines, Stirling engines, steam turbines, fuel cells, and solar panels. The components of the subsystem include absorption chillers, adsorption chillers, and dehumidifiers for cooling purposes.

Wang et al. [8] proposed a new combined cooling and power system based on LiBr/H$_2$O cycle. Figure 4.8 portrays the system configuration. An ejector is used to reduce the turbine back pressure to facilitate performance improvement during power generation. The proposed system consists of a rectification column, a pre-heater, a solution heat exchanger, pumps, valves, an absorber, an ejector, a turbine, a superheater, a condenser, a cooler, and an evaporator.

Pure ammonia is produced from the rectification column. One stream of the ammonia from the rectification column is cycled to the superheater for power generation. Another stream of ammonia is condensed into liquid in the condenser. One stream of the condensed ammonia flows back into the rectification column. Another stream of condensed ammonia is cooled and sprayed in the evaporator for cooling production. The high-temperature heat transfer fluid flows through the superheater, rectification column, and preheater successively. In this way, thermal energy is recovered to generate electricity and cooling.

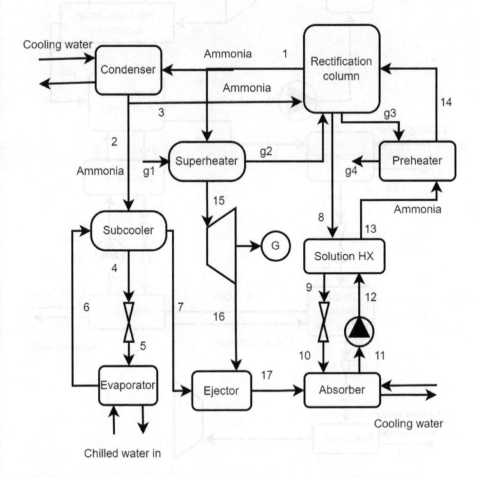

**FIGURE 4.8**   Schematical diagram of the CCP system configuration[8].

### 4.1.2.2 CHP Systems

Combined heating and power (CHP) systems typically consist of a prime mover working in tandem with a waste heat recovery subsystem. The prime mover utilizes fuel to generate electricity. The heat recovery subsystem harvests the dissipated waste heat from the prime mover to produce useful heating. The subsystem includes several combinations of heat exchangers and heat pumps for heating purposes.

### 4.1.2.3 CCHP Systems

Zoghi et al. [9] proposed a cascade organic quadrilateral cycle LiBr/$H_2$O cycle-compression system. Figure 4.9 portrays the configuration of the combined cooling,

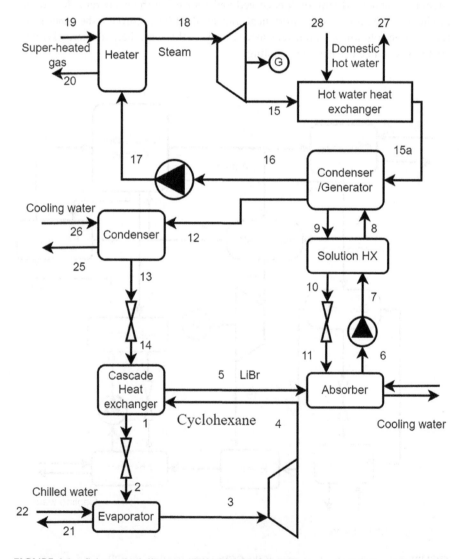

**FIGURE 4.9** Schematical diagram of the CCHP system [9].

heating, and power (CCHP) system. The proposed CCHP system comprises a heater, an expander, domestic hot water exchangers, condensers, LiBr/H$_2$O cycle, and a compression cycle. Cyclohexane is employed as the working fluid for the organic quadrilateral cycle.

By harnessing the thermal waste heat from the exhausted gas, the cyclohexane vapor is generated and allowed to flow into the expander to produce power. After that, the exhausted vapor is used to heat the water for domestic utilization. Subsequently, the superheated vapor is employed as the heat source to operate the absorption chiller. LiBr/H$_2$O and R140 are the working fluids of the absorption chiller and compression cycle, respectively. Key results from the research have shown that the propylene glycol-water mixture can be cooled down from −5 to −15°C. Thus, the proposed cascade organic quadrilateral cycle LiBr/H$_2$O cycle-compression system is able to utilize thermal waste heat in order to produce electricity, heating, and cooling.

### 4.1.2.4   CCHPP Systems

Multigeneration systems offering superior energy utilization efficiency are considered to be the next alternatives to conventional energy supply methods. Chen and Chua [10] proposed a combined cooling, heating, potable water, and power system (CCHPP). Figure 4.10a portrays the schematical diagram of the CCHPP system. The proposed CCHPP consists of a gas turbine, an absorption chiller, a hot water tank, and an adsorption chiller. Compressed natural gas is employed as the feeding fuel for the gas turbines to produce electricity. Cooling water is used to harness the waste heat from the super-heated exhaust. The heated water is used to drive the absorption chiller to produce cooling. The outlet hot water from the absorption chiller is then employed to run the adsorption chiller. The adsorption chiller is able to simultaneously produce chilled water and potable water. Further, the outlet hot water from the adsorption chiller is employed to regenerate solid desiccant-coated heat exchangers used for the purpose of air dehumidification. Accordingly, dry and cool air is produced. In this manner, the proposed CCHPP system utilizes natural gas as the main fuel to produce electricity, cooling, heating, and potable water. Figure 4.10b presents the energy flow of the CCHPP system under design conditions.

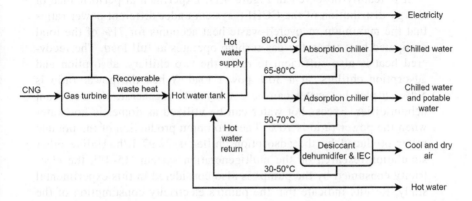

**FIGURE 4.10**   Schematic of the proposed CCHPP system [10].

In this section, a novel combined cooling, heating, power, and potable water system is proposed and experimentally investigated. The proposed system's performance is firstly assessed by means of a 4E (exergy, energy, economic, and environment) off-design analytical method. The employed gas turbine is introduced in Chapter 1, Section 1.1. The deployed absorption chiller, adsorption chiller, and desiccant dehumidifier have been presented in the previous chapters.

a. Energy-saving performance analysis

Primary energy-saving ratio (PESR) is an essential indicator to evaluate the primary energy-saving performance of the proposed CCHPP system compared to an equivalent conventional system [11]. It is defined as:

$$\text{PESR} = 1 - \frac{E_{gt}\eta_{grid}}{E_{grid}\eta_{gt}} \tag{4.41}$$

$$E_{grid} = E_{gt} + \frac{(Q_{c,ab} + Q_{c,ad})}{COP_{ec}} + \frac{Q_{hw}}{COP_{eb}} + E_{ro} + W_{pw} \tag{4.42}$$

where $E_{gt}$ and $E_{grid}$ represent the electricity produced from the gas turbine and the utility grid. $\eta$ denotes the efficiency. A national grid efficiency $\eta_{grid}$ of 0.33 [12] is adopted in the present analysis. The electric chiller's coefficient of performance ($COP_{ec}$) is designated to be 3.0 [13]. $W_{pw}$ is the electricity consumption of the RO desalination unit. The energy consumption of the small-scale RO system is 15 kWh/m³ [14]. The proposed system's $COP_{sys}$ is expressed as:

$$COP_{sys} = \frac{E_{gt} + Q_{c,ab} + Q_{c,ad} + Q_{c,dd} + Q_{hw}}{\dfrac{E_{gt}}{\eta_{gt}} + E_p} \tag{4.43}$$

where $E_p$ represents the total electricity consumption of the pumps in the system. $Q$ denotes the produced cooling or heating.

It is readily observed in Figure 4.11. Experimental performance of energy distribution of the CCHPP system under different power ratios that the maximum recovered waste heat accounts for 71% of the total dissipated heat when the gas turbine operates at full load. The recovered heat is also sufficient to drive the two chillers, absorption and adsorption chillers, at a 30% power load. When the power ratio is 50%, surplus hot water can be employed to regenerate the desiccant dehumidifier. Excess hot water can be utilized as domestic hot water when the power ratio is 70%. The maximum production of the potable water produced by the adsorption chiller is 22.97 L/h. Unlike other simulation research on the multigeneration system [15–18], the electricity consumed by the pumps is also considered in this experimental study. Results indicate that the pump's electricity consumption of the proposed system accounts for around 2%–6% of the total energy. It is

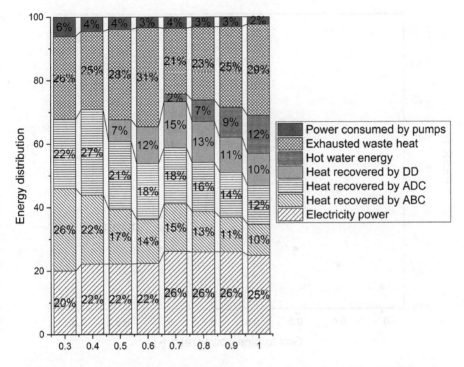

**FIGURE 4.11** Experimental performance of energy distribution of the CCHPP system under different power ratios.

noteworthy that the different nominal cooling capacity combinations of the three subsystems can potentially affect the energy-saving performance of the whole system, as the COPs of the subsystems are different when utilizing waste heat to produce cooling. However, the optimization of the subsystem's cooling capacity combination is not within the scope of this study.

It is observed from Figure 4.12 that the maximum PESR is 32.5% when the system operates at 100% load. The COP of the proposed system also appreciates from 0.54 to 0.62 when the power ratio is regulated from 30% to 100%. That observation is attributed to the fact that the efficiency of the gas turbine increases with the higher power ratio. As a result, more generated heat is recovered. When the power ratio is below 40%, it is not suggested to operate the multigeneration system since the PESR is negative. Figure 4.13 portrays the energy flow of the system under design conditions. If all the recovered waste heat is converted to hot water without the production of chilled water, the total efficiency of the proposed system is 0.80. The total cooling efficiency of employing hot water to produce cooling is 0.49. Unlike many of the simulation reported in the literature that ignores the effect of friction and heat transfer to the ambient when evaluating the system performance, the experimental result indicates that the energy loss to produce chilled water is 3 kW.

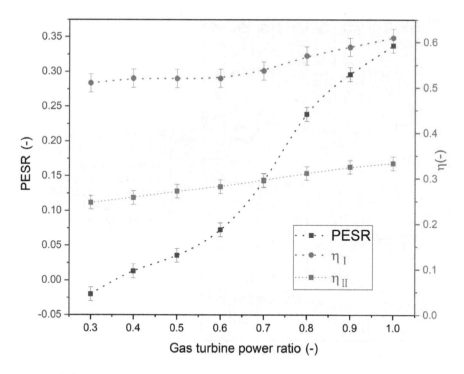

**FIGURE 4.12** Experimental performance of the PESR and COP performance of the proposed CCHPP system under different power ratios.

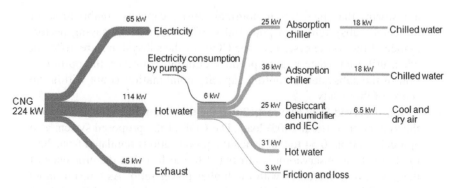

**FIGURE 4.13** Sankey diagram of energy flow in the proposed system under design conditions.

b. Exergy analysis

Exergy analysis is known to be a useful and practical tool to quantify energy loss and locate the associated exergy destruction [19]. The kinetic and potential exergy of the system is typically ignored. Thus, the exergy (*ex*) consists of both physical exergy and chemical exergy. The physical exergy denotes the maximum useful work with respect to the ambient conditions (at 25°C and 1 atm). The chemical exergy is described as the leaving composition of a stream from its equilibrium of the ambient [15]. The exergy balance equation is presented as follows according to Dinçer and Rosen [20]:

$$\dot{Ex}_Q + \sum_m \dot{m}_{in}ex_{in} = \sum_n \dot{m}_{out}ex_{out} + \dot{Ex}_w + \dot{Ex}_D \tag{4.44}$$

$$ex = ex_{ph} + ex_{ch} \tag{4.45}$$

$$\dot{Ex}_Q = \left(1 - \frac{T_o}{T_i}\right)\dot{Q}_i \tag{4.46}$$

$$\dot{Ex}_w = \dot{W} \tag{4.47}$$

where $\dot{Ex}_Q$ is the exergy rate caused via heat transfer. $\dot{Ex}_w$ and $\dot{Ex}_D$ represent the exergy rate caused by work and destruction, respectively. The chemical exergy is described as follows according to Bejan et al. [19]:

$$ex_{ch,mix} = \sum_j^k X_j ex_{ch,j} + RT_0 \sum_{j=1}^k X_j \ln(X_j) \tag{4.48}$$

The specific exergy of the cooling water, hot water, or air stream is defined as:

$$ex_{ph} = (h - h_0) - T_0(s - s_0) \tag{4.49}$$

Exergy efficiency is defined as the ratio of the total product's exergy to the fuel exergy and is written as:

$$\eta_{ex,cchpp} = \frac{\dot{W}_{net} + \dot{Ex}_h + \dot{Ex}_{cooling}}{\dot{Ex}_f} = 1 - \frac{\dot{E}_D}{\dot{Ex}_f} \tag{4.50}$$

$$\dot{Ex}_f = \dot{Ex}_{f,ph} + \dot{Ex}_{f,ch} \tag{4.51}$$

$$\dot{Ex}_{f,ph} = \dot{m}_f c_p \left(T - T_0 - T_0 \ln\frac{T}{T_0}\right) \tag{4.52}$$

$$\dot{Ex}_{f,ch} = \dot{m}_f \left(RT_0 \sum_i x_i \ln x_i + \sum_i x_i ex_{ch}^i\right) \tag{4.53}$$

where $c_p$ is the specific heat capacity, $R$ represents the universal gas constant. $x_i$ and $ex_{ch}^i$ denote the mole fraction and standard chemical exergy of the $i$th components. The relative values are adopted from the literature [21]. In addition, the water vapor and dry air are considered ideal gas in the desiccant dehumidifier, indirect evaporative cooler, and cooling tower during

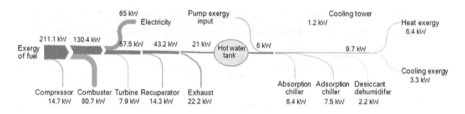

**FIGURE 4.14** Sankey diagram of the exergy flow of the proposed system under design conditions.

the exergy analysis process. The specific exergy of the air is described as follows according to Avanessian and Ameri [22]:

$$ex_{air} = (C_{p,da} + \omega C_{p,v})\left(T - T_0 - T_0 \ln \frac{T}{T_0}\right) + R_{da}T_0(1 + 1.6078\omega)\ln \frac{P}{P_0}$$

$$+ R_{da}T_0(1 + 1.6078\omega)\ln \frac{(1 + 1.6078\omega_0)}{(1 + 1.6078\omega)} + 1.6078\omega \ln \frac{\omega}{\omega_0}$$

(4.54)

The variation of the exergy efficiency with respect to different power ratios is depicted in Figure 4.12. The maximum exergy efficiency of the proposed system is 0.32 under design conditions. Figure 4.14 illustrates the exergy flow of the proposed multigeneration system under design conditions. It is observed that the exergy efficiency of the gas turbine is 0.29. When the waste heat is fully recovered to generate hot water without producing cooling, the exergy efficiency approaches 0.41. The exergy efficiency of employing thermal heat embedded in hot water to produce cooling is 0.16. It is noteworthy that the calculated second-law efficiency of the chillers is lower than some published values [23,24]. This is because these derived values from the literature are based on simulations that adopt many assumptions, neglecting pump exergy input and friction loss. It is also observed from Figure 4.14 that the exergy destruction rate is significantly high at the combustor of the gas turbine. This is because the power production (electricity generation) process is accompanied by a significant exergy supply and destruction [25]. The exergy destruction rate of the waste heat recovery subsystems is much lower than the gas turbine, accounting for 11% of the total exergy destruction rate. The cooling capacities of the adsorption chiller and absorption chiller are similar, while the exergy destruction of the adsorption chiller is comparatively higher than the absorption chiller. The results indicate that the absorption chiller's performance is superior to the adsorption chiller from both the first and second law perspectives.

c. Economic analysis

A multigeneration system is known to incur a relatively higher initial investment but a lower operating cost when compared with conventional separated systems. Therefore, it is of great significance to compute the time required to recover the proposed CCHPP system's capital cost. The payback period (PBP) is employed to evaluate the economic feasibility of the proposed CCHPP system and is expressed as follows according to Kong et al. [26]:

$$PBP = \frac{TII}{AOCS} \tag{4.55}$$

$$TII = \sum_i Pr_i Q_{rated,i} + IC + CC + MC \tag{4.56}$$

$$AOCS = COST_{sp} - COST_{cchpp} \tag{4.57}$$

$$COST_{sp} = \left( \frac{E + \dfrac{Q_h}{\eta_{eb}} + \dfrac{Q_c}{\eta_{ec}} + Q_w}{\eta_{grid}} \right) * Pr_f * t \tag{4.58}$$

$$COST_{cchpp} = Pr_f * F * t + Pr_e * E_p * t \tag{4.59}$$

where TII and AOCS represent the total initial investment and the annual operational cost saving. $Pr_i$ denotes the unit price, and $Q_{rated,i}$ represent the rated power of the $i$th component of the proposed system. The related prices are tabulated in Table 4.3. These values have been commonly adopted in the literature in recent years. The actual price of deploying the proposed system in different countries may vary slightly. IC is the installation cost. CC is the cost of the control system, and MC is the maintenance cost.

The economic assessment of the proposed multigeneration system is illustrated in Figure 4.15. It is observed from Figure 4.15 that both a higher annual operation time and a higher power ratio yield higher annual profit. The maximum annual profit peaks at 80,000 USD/year. There are works that calculate payback year based on specific operating conditions [15,17,29]. The present system's payback year is computed with varying power ratios and annual operation time. Considering the gas turbine's maintenance time and

## TABLE 4.3
## Commonly Adopted Price of Different Loads

| Product | Annual Average Price | Unit | References |
|---|---|---|---|
| Power | 0.155 | $/kWh | [27] |
| Heating | 0.023 | $/kWh | [27] |
| Cooling | 0.07 | $/kWh | [27] |
| Fuel | 0.5 | $/m³ | [18] |
| Potable water | 0.986 | $/m³ | [28] |

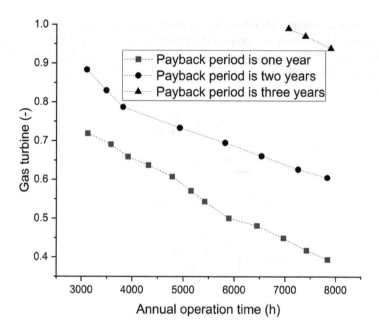

**FIGURE 4.15** Payback period of the system operating under different power load ratios and annual operation hours.

breakdown period, the annual operation time is unable to reach 8,760 hours. In addition, the average operation ratio and annual operation time should be no less than 0.7 and 7,000 hours, respectively, in order to achieve decent economic viability. The payback period performance map further provides a fast and convenient means to assess the economic performance of the proposed system before installation. Once the average power load and operation time on the consumer side is estimated based on the demands, the payback period of the proposed system is estimated. For instance, if the annual operation time is 7,500 hours and the power ratio is 1.0, the payback period is obtained to be 0.7 years. The payback year becomes 2.1 years when the average annual operation time and the power ratio are 7,000 hours and 0.7, respectively.

d. Environmental analysis

Global warming is an exacting issue that needs addressing to promote environmental preservation. Carbon dioxide emission reduction ratio (CDERR) and annual $CO_2$ emission reduction (CDER) are two key indicators commonly used to evaluate the environmental performance of multigeneration systems [30]. They are expressed as:

$$CDERR = \frac{CDER}{CDE_{sp}} \times 100\% \qquad (4.60)$$

$$CDER = CDE_{sp} - CDE_{cchpp} \qquad (4.61)$$

where $CDE_{sp}$ and $CDE_{cchpp}$ can be calculated as:

$$CDE_{cchpp} = \mu_f F \tag{4.62}$$

$$CDE_{sp} = \mu_e \left( E + \frac{Q_h}{\eta_{eb}} + \frac{Q_c}{\eta_{ec}} + Q_w \right) \tag{4.63}$$

where $\mu$ is the average emission factor. It is noteworthy that the emission factor evaluates the effects of both $CO_2$ and other polluting ($NO_x$, $SO_x$, and particles) [31] (Table 4.4).

Reducing $CO_2$ emission is a key performance indicator towards carbon neutrality [36]. The effects of the power load ratio and annual operation time on the carbon dioxide emission reduction (CDER) and carbon dioxide emission reduction ratio (CDERR) performance of the proposed system are depicted in Figure 4.16. It is observed from Figure 4.16 that

## TABLE 4.4
## Price of Different Types of Equipment for the Proposed System

| Components | Initial Investment | Unit | References |
|---|---|---|---|
| Gas turbine | 856.5 | $/kW | [27] |
| Absorption chiller | 165.5 | $/kW | [27] |
| Adsorption chiller | $557*Q_{ad}$ | $/kW | [32] |
| Desiccant dehumidifier | $2*130*(A/0.093)^{0.78}$ | $ | [33] |
| Water tank | $82.3*V_{wt} + 179$ | $ | [32] |
| Cooling tower | $54.8*m_{ct} + 1412$ | $ | [32] |
| Electric chiller | 153 | $/kW | [34] |
| Electric boiler | 46.5 | $/kW | [35] |
| Pump | $3540*(W_p)^{0.71}$ | $ | [16] |
| Control system | 5% of capital cost | – | |
| Maintenance cost of the system | 1.25% of capital cost | – | [32] |
| Installation cost of the system | 5% of capital cost | – | [32] |

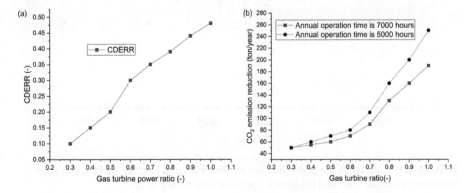

FIGURE 4.16　(a) CDERR; and (b) CDER performance of the proposed system.

the gas turbine's increase in power load contributes to attaining better CDER and CDERR. This is because a higher power load leads to improved energy utilization performance. Consequently, a lesser amount of $CO_2$ is emitted. In addition, a higher annual operation time also facilitates better CDER performance. In comparison, the CDERR only varies with the power load ratio. As the load ratio varies from 0.3 to 1.0, CDERR appreciates from 0.02 to 0.48. It is apparent from Figure 4.16b that the maximum $CO_2$ emission reduction is 280 tons/year when compared with the conventional energy supply system. Accordingly, the gas turbine ought to operate at higher loading conditions to arrive at better CDERR.

e. Effect of the national grid efficiency

Figure 4.17 depicts the variation of PESR and payback period with respect to $\eta_{grid}$ when the gas turbine operates under full load with an annual operation time of 7,000 hours. The proposed system yields a PESR of 0.24 when the $\eta_{grid}$ reaches 0.38. In addition, the payback period is 3.2 years. These data highlight the capacity of the proposed system of achieving excellent energy-saving performance. The results also support the fact that the proposed system gains more advantages when installed in areas where $\eta_{grid}$ is low. For instance, when the $\eta_{grid}$ is 0.3, the PESR and payback years yield 0.39 and 1.5, respectively. In areas that are far away from a power plant, the

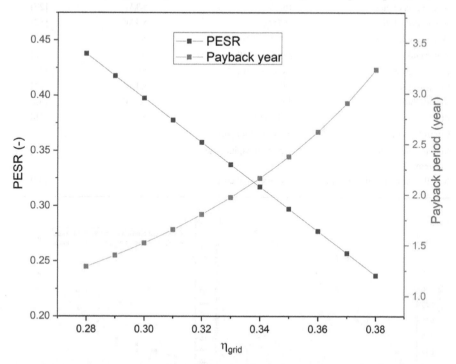

**FIGURE 4.17** Effect of the grid efficiency on the PESR and payback period. The gas turbine operates at full load, and the annual operation time is 7000 hours.

grid efficiency is relatively low due to the energy loss during power transmission. In this regard, the proposed system is highly advantageous to be deployed as a distributed energy system in place of the conventional energy supply system.

f. Effect of the fuel price

It is a commonplace knowledge that the natural gas price for households varies in different countries [37]. It is, therefore, essential to investigate the effect of a fluctuating fuel price on the deployment of the proposed multigeneration system. Results in Figure 4.18 convey that the payback year of the proposed system is significantly affected by the annual operation hours and fuel price. As fuel price appreciates, from 0.02 to 0.14 USD/kWh, the payback period is shortened from 3.8 to 0.7 years when the annual operation time is 7,000 hours. The result implies that the proposed system ought to be deployed in areas experiencing high fuel prices. It is also observed that the annual operation time affects the payback year of the proposed system. For instance, when the fuel price is 0.08 USD/kWh, the annual operation time has to be more than 7,000 hours to achieve a payback period of less than one year. Comparatively, when the fuel price is 0.06 USD/kWh, the annual operation time ought to be longer than 8,000 hours to reap a payback year of less than one year. Thus, both annual operation time and fuel price are factors that must be seriously considered when deploying the proposed multigeneration system.

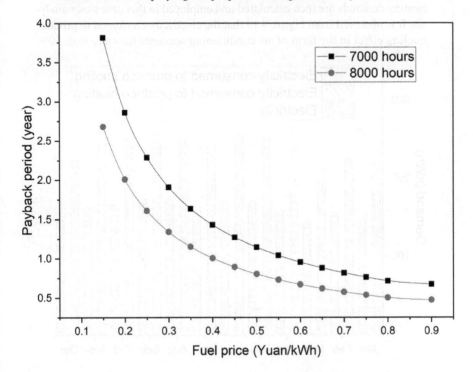

**FIGURE 4.18** Variation of payback year of the proposed system with respect to fuel price; The gas turbine operates under full load, and the national grid efficiency is 0.33.

Key results reveal that 71% of the dissipated waste heat can be recovered, and the primary energy-saving ratio is 32.5%. In addition, the proposed system is capable of reducing up to 256 tons of $CO_2$ emission every year compared with the conventional system. Results further highlight that the payback period of the proposed system is about 1.1 years when the annual operation time and power ratio of the gas turbine are 8,000 h and 0.9, respectively. In addition, the effects of different variables on the economic feasibility of the proposed system are analyzed. Results also indicate that both national grid efficiency and fuel price exert a greater influence on the payback year of the proposed system. Further, a case study is performed based on a typical tropic city's electricity, cooling, and heating demands. Lastly, the proposed system can be employed as a promising energy source supply taking into consideration its excellent energy-saving potential, environmental sustainability, and economic feasibility.

g. Case study

In this section, a case study is adopted for the purpose of investigating the 4E performance of the proposed system under practical conditions. Figure 4.19 shows the average monthly household electricity consumption of a five-room apartment in Singapore in 2019. Xu and Ang [38] provided a detailed statistical analysis of the daily electricity consumption distribution of different housing types in Singapore. The average electricity, cooling, and heating demands are then calculated and employed in this case study analysis. It is observed from Figure 4.19 that the electricity consumed to produce cooling effect in the form of air-conditioning accounts for more than 50%.

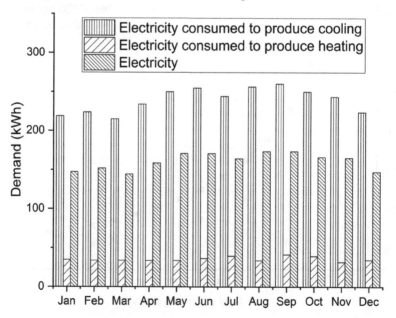

**FIGURE 4.19** Average monthly household demand consumption by dwelling type (five-room and executive) in Singapore in 2019 [39].

It is also noteworthy that the demand variations among different months are marginal. This is because Singapore is a tropical country residing on the equatorial belt, with a relatively stable ambient temperature throughout the year [38]. In this study, the proposed capacity of the multigeneration system is sized to provide electricity, cooling, potable water, and heating for 123 five-room apartments in Singapore. It is assumed that one person per day consumes 3 L of potable water. Conventionally, the cooling and heating demands are supplied by electrical chillers and boilers. The consumed electricity is directly obtained and purchased from the grid. The average monthly electricity and gas tariffs are illustrated in Figure 4.20.

The proposed system operates at its design conditions with an annual operation time of 7,500 hours. As shown in Figure 4.20, a part of the cooling demand is provided by the heat-driven chillers utilizing the available waste heat. The inadequate part of cooling demand is then compensated by the electric chiller. It can be observed from Figure 4.21 that the heating produced by the proposed system exceeds the consumer's heating demand. In other words, the recoverable waste heat is not sufficiently utilized. The results indicate that enhancing the heat-driven chillers' cooling capacities contributes significantly to improving the energy utilization efficiency of the proposed system. In addition, the proposed multigeneration system is able to produce about one-third of the consumers' potable water needs. As shown in Table 4.5, the payback period and the annual profits are 1.98 years and 44,920 $/year, respectively. These results show

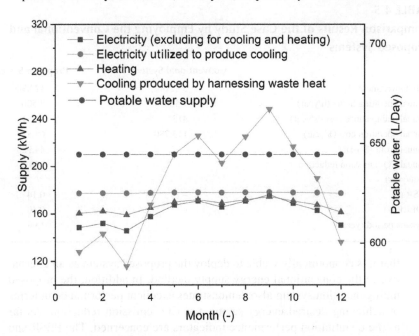

**FIGURE 4.20** Monthly electricity, heating, cooling, and potable water supply by the proposed CCHPP system in the case study.

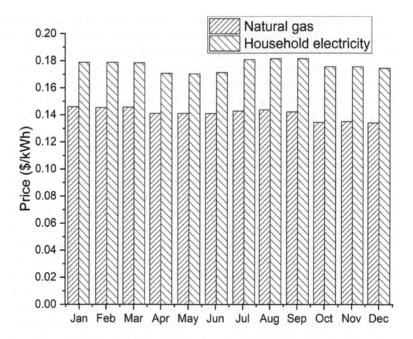

**FIGURE 4.21** Average monthly electricity and gas tariffs in Singapore in 2019 [40].

**TABLE 4.5**

**Comparison Results of the Case Study by Employing the Conventional and Proposed Systems**

|  | Conventional System | Proposed System |
|---|---|---|
| Initial investment ($) | 35,214 | 123,980 |
| Annual operating hours (h/year) | 7,500 | 7,500 |
| Annual maintenance cost ($/year) | 413 | 1,261 |
| Annual operation cost ($/year) | 113,780 | 68,859 |
| Annual profit ($/year) | – | 44,920 |
| Annual $CO_2$ emission reduction (ton/year) | – | 248 |
| PESR (-) | – | 0.14 |
| CDERR (-) | – | 0.40 |
| Payback period (years) | – | 1.98 |

that it is economically viable to deploy the proposed system as an alternative to the conventional energy supply method. In addition, the proposed multigeneration system also demonstrates excellent performance in terms of achieving desired energy-saving and $CO_2$ emission reduction. As far as the quantitation performance indicators are concerned, The PESR and CDERR are 0.14 and 0.40, respectively. Compared with the conventional system, the proposed system is estimated to be able to reduce 248 tons of $CO_2$ emissions every year.

This works entails the 4E performance assessment of a novel multigeneration system comprising two heat-driven chillers and several experimental heat recovery subsystems. The proposed system is able to simultaneously provide four key utilities, namely, electricity, cooling, heating, and potable water, via a cascading heat recovery process. The proposed system's performance is comprehensively investigated via an off-design performance analysis method. Further, a case study is adopted to demonstrate with physical data that the proposed multigeneration system is able to achieve excellent energy-saving, environmental sustainability, and economic feasibility under practical conditions. Key results reveal that 71% of the dissipated waste heat can be recovered, and the primary energy-saving ratio is 32.5%. In addition, the proposed system is capable of reducing up to 280 tons of $CO_2$ emission every year compared with the conventional system. Results further highlight that the payback period of the proposed system is about 1.1 years when the annual operation time and power ratio of the gas turbine are 8,000 hours and 0.9, respectively. In addition, the effects of different variables on the economic feasibility of the proposed system are analyzed. Results also indicate that both national grid efficiency and fuel price exert greater influence on the payback year of the proposed system. Further, a case study is performed based on a typical tropic city's electricity, cooling, and heating demands. Lastly, the proposed system can be employed as a promising energy source supply taking into consideration its excellent energy-saving potential, environmental sustainability, and economic feasibility.

## 4.2 SYSTEM APPLICATIONS ON WASTE COLD ENERGY RECOVERY

### 4.2.1 Integrated Systems with Single Products

#### 4.2.1.1 Combined Cycle System (Rankine Cycle + Direct Expansion)

It was earlier mentioned in Chapter 3.2 that the cold energy utilization efficiency is relatively low when only one cycle is adopted for power generation. The configuration that integrates the Rankine cycle and direct expansion is the most common combined cycle that has been adopted during LNG cold energy power generation. The schematic diagram and T-Q diagram of the combined cycle is shown in Figure 4.22a and b, respectively. LNG from the storage tank is first pumped to a pressure higher than the distribution network. The high-pressure LNG is then sent to the Rankine cycle and employed as a heat sink. Finally, the natural gas from the Rankine cycle is heated and expanded in the direct expansion turbine. Conventionally, seawater is adopted as the heat source since the receiving terminal is located usually adjacent to the port area. Compared to either a pure Rankine or direct expansion cycle, the employment of two turbines instead of one can produce a greater amount of power in the combined cycle system. The cold energy recovery rate is defined as:

$$\text{CRR} = \frac{W_{t,lng} + W_{t,\text{rankine}} - W_{p,lng} - W_{p,\text{rankine}}}{m_{lng}(h_{10} - h_5)} \tag{4.41}$$

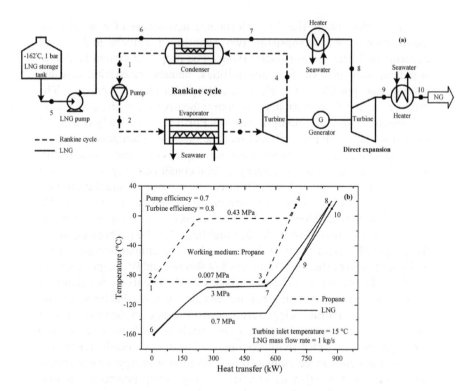

**FIGURE 4.22** Combined cycle for LNG cold energy utilization: (a) Schematic diagram, (b) T-Q diagram.

**TABLE 4.6**
**Detailed Thermodynamic Parameters of the Combined Cycle**

| Points | Fluid | m (kg/s) | p (MPa) | T (°C) | h (kJ/kg) | s (kJ/kg.K) |
|--------|-------|----------|---------|--------|-----------|-------------|
| 1 | Propane | 1.16 | 0.01 | −88.90 | −0.60 | 0.12 |
| 2 | Propane | 1.16 | 0.43 | −88.66 | 0.36 | 0.12 |
| 3 | Propane | 1.16 | 0.43 | 15.00 | 602.71 | 2.49 |
| 4 | Propane | 1.16 | 0.01 | −88.90 | 469.15 | 2.67 |
| 5 | LNG | 1.00 | 0.12 | −162.00 | −1.77 | −0.02 |
| 6 | LNG | 1.00 | 3.00 | −160.41 | 7.93 | 0.01 |
| 7 | LNG | 1.00 | 3.00 | −93.90 | 551.22 | 3.40 |
| 8 | LNG | 1.00 | 3.00 | 15.00 | 856.91 | 4.76 |
| 9 | LNG | 1.00 | 0.70 | −58.42 | 719.16 | 4.93 |
| 10 | LNG | 1.00 | 0.70 | 10.00 | 870.27 | 5.54 |
| Overall | CRR | 31.92% | | | | |

The detailed thermodynamic parameters of the combined cycle are listed in Table 4.6. According to the obtained results, the theoretical cold energy recovery rate peaks at 31.92%. It is worth noting that the LNG regasification pressure has a combined effect on the overall cycle efficiency. In other words, a higher regasification pressure adversely impacts the efficiency of the Rankine cycle while positively influencing the power output of the direct expansion turbine. The impact of the LNG regasification pressure on the combined cycle efficiency is highlighted in Figure 4.23a. If the condensing pressure of the Rankine cycle remains constant, efficiency degradation

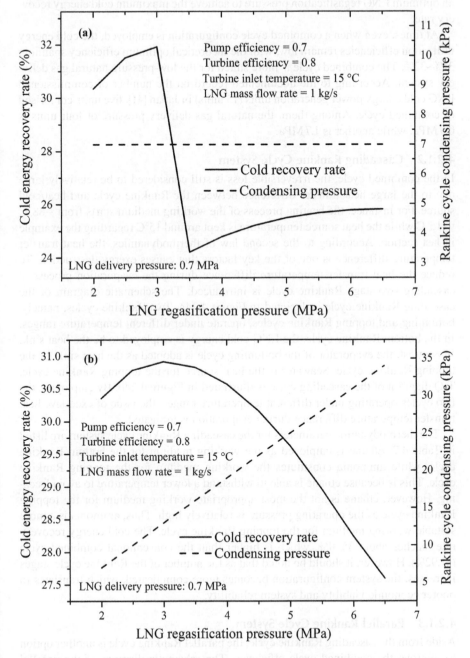

**FIGURE 4.23** Impact of LNG regasification pressure on the combined cycle efficiency: (a) constant condensing pressure, (b) varying condensing pressure.

still exists. This is because once the LNG evaporating temperature exceeds the condensing temperature of the Ranking cycle, the latent heat can no longer be utilized. If the condensing pressure varies with the LNG regasification pressure, the efficiency degradation can be avoided, as shown in Figure 4.23b. It is apparent that there exists an optimum LNG regasification pressure to achieve the maximum cold energy recovery rate.

At times, even when a combined cycle configuration is employed, the cold energy utilization efficiencies remain low, with the theoretical optimum efficiency spanning 31%~32%. The combined cycle is more suited for the low-pressure natural gas delivery system. According to the documented record on the number of commissioned LNG cold energy power generation units (15 units) in Japan [41], five units comprised of combined cycle. Among them, the natural gas delivery pressure of four units is 0.9 MPa, while another is 1.7 MPa.

### 4.2.1.2   Cascading Rankine Cycle System

In the combined cycle, the irreversible loss is still considered to be relatively large due to the large heat transfer difference between the Rankine cycle and heat/cold source. For instance, the heating process of the working medium starts from −88°C to 15°C while the heat source temperature is kept around 15°C regarding the example in last section. According to the second law of thermodynamics, the heat transfer temperature difference is one of the key factors that causes exergy destruction. To reduce the heat transfer temperature difference during the evaporation process, a cascading two-stage Rankine cycle is introduced. The schematic diagram of the cascading Rankine cycle is depicted in Figure 4.24a. Two Rankine cycles, namely, bottoming and topping Rankine cycles, operate under different temperature ranges. In the bottom Rankine cycle, the LNG cold energy is employed to be the heat sink. In contrast, the evaporator of the bottoming cycle is adopted as the heat sink for the topping Rankine cycle. Seawater is the heat source for the topping Rankine cycle. T-Q diagram of the cascading cycle is illustrated in Figure 4.24b. By employing the two cycles operating under different temperature ranges, the issue of excessive heat transfer temperature difference during evaporation is mitigated.

The thermodynamic parameters for the cascading Rankine cycle system are listed in Table 4.7. Ethane is employed as the working medium for the bottom Rankine cycle, while ammonia constitutes the working medium for the topping Rankine cycle. This is because ethane is able to withstand a lower temperature to avoid freezing. However, ethane is not the most appropriate working medium for the topping Rankine cycle as the operating pressure is relatively high. Thus, ammonia is a more suitable working medium for the topping Rankine cycle. The cold energy recovery rate reaches about 35.91%, which is higher than the conventional combined cycle (31.92%). However, it should be noted that as the number of the Rankine cycle stages increases, the system configuration becomes more complicated, which translates to poorer economic viability and system reliability.

### 4.2.1.3   Parallel Rankine Cycle System

Aside from the cascading Rankine cycle, the parallel Rankine cycle is another option to improve the combined cycle efficiency. The schematic diagram of the parallel

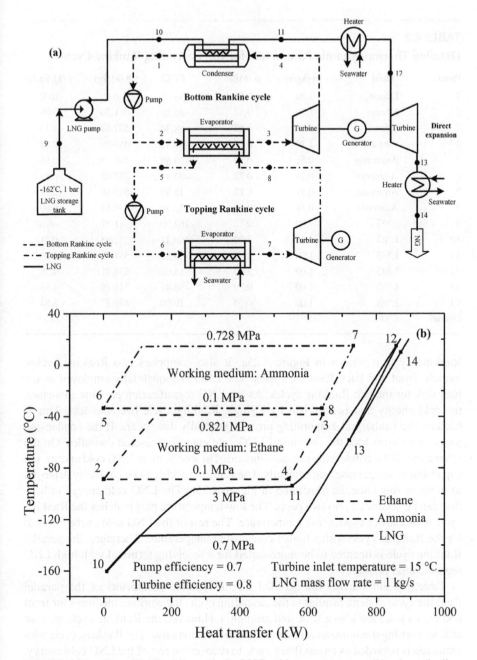

**FIGURE 4.24** Cascading Rankine cycle for LNG cold energy recovery: (a) schematic diagram, (b) T-Q diagram.

**TABLE 4.7**

**Detailed Thermodynamic Parameters for the Cascading Rankine Cycle**

| Points | Fluid | $m$ (kg/s) | $p$ (MPa) | $T$ (°C) | $h$ (kJ/kg) | $s$ (kJ/kg.K) |
|---|---|---|---|---|---|---|
| 1 | Ethane | 1.20 | 0.10 | −88.82 | −0.60 | 0.00 |
| 2 | Ethane | 1.20 | 0.82 | −88.36 | 1.30 | 0.00 |
| 3 | Ethane | 1.20 | 0.82 | −38.30 | 537.42 | 2.35 |
| 4 | Ethane | 1.20 | 0.10 | −88.82 | 453.98 | 2.65 |
| 5 | Ammonia | 0.51 | 0.10 | −33.59 | 190.75 | 0.88 |
| 6 | Ammonia | 0.51 | 0.73 | −33.41 | 192.07 | 0.88 |
| 7 | Ammonia | 0.51 | 0.73 | 15.00 | 1619.58 | 5.91 |
| 8 | Ammonia | 0.51 | 0.10 | −33.59 | 1440.14 | 6.60 |
| 9 | LNG | 1.00 | 0.12 | −162.00 | −1.77 | −0.02 |
| 10 | LNG | 1.00 | 3.00 | −160.41 | 7.93 | 0.01 |
| 11 | LNG | 1.00 | 3.00 | −93.82 | 551.59 | 3.40 |
| 12 | LNG | 1.00 | 3.00 | 15.00 | 856.91 | 4.76 |
| 13 | LNG | 1.00 | 0.70 | −58.42 | 719.16 | 4.93 |
| 14 | LNG | 1.00 | 0.70 | 10.00 | 870.27 | 5.54 |
| Overall | CRR | 35.91% | | | | |

Rankine cycle is shown in Figure 4.25a. It also comprises two Rankine cycles, namely, front and back Rankine cycles. The LNG is sequentially employed as the heat sink for the two Rankine cycles. As the LNG regasification pressure increases, the cold energy that is absorbed by a single Rankine cycle becomes less. This is because the isothermal evaporating process gradually disappears. If the condensing pressure remains low, the amount of LNG cold energy recovered dwindles. On the other hand, if the condensing pressure increases to absorb more LNG cold energy, the expansion ratio decreases significantly. The Rankine cycle can be parallelly deployed to mitigate this issue, as illustrated in Figure 4.25b. The LNG cold energy utilization can be separated into two parts. The low-temperature part matches the Rankine cycle with a lower condensing temperature. The rest of the LNG cold energy is used for the Rankine cycle with a higher condensing temperature. Therefore, the parallel Rankine cycle is deemed to be more suited for a receiving terminal with high LNG regasification pressure.

Table 4.8 summarizes the detailed thermodynamic parameters of the parallel Rankine cycle system. Similar to the cascading cycle, the working mediums for front and back cycles are also ethane and ammonia. However, the Rankine cycle with an ethane working medium operates at a higher expansion ratio. The Rankine cycle with ammonia is regarded as an auxiliary cycle to recover the rest of the LNG cold energy. Extensive calculations reveal that the cold energy recovery rate approaches 32.82%.

### 4.2.1.4  Cold Energy Storage-Based System

Although high-efficiency power generation cycles can improve cold energy utilization rate, their capacities are still constrained by the minimum LNG regasification rate. LNG receiving terminals are mainly built to meet the natural gas demand

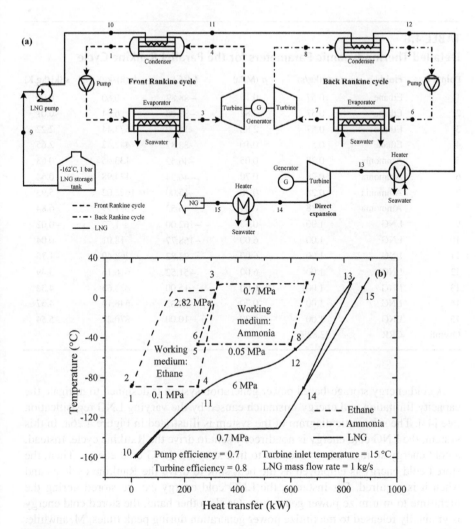

**FIGURE 4.25** Parallel Rankine cycle for LNG cold energy recovery: (a) schematic diagram, (b) T-Q diagram.

from power plants. The natural gas demand from these sectors usually appears under seasonal or hourly variations. If the demand profile of the cold energy users is a stable one, then the capacity of the cold energy utilization technologies is designed to be 20%–30% of their baseload [42]. In addition, the energy mismatch between systems and users is also an important issue. For instance, the electricity produced by the LNG cold energy power generation system can, ironically, become a burden for the national grid during the off-peak period. Therefore, in addition to merely improving the system efficiency, the energy mismatch issue should be judiciously resolved.

**TABLE 4.8**

**Detailed Thermodynamic Parameters for the Parallel Rankine Cycle**

| Points | Fluid | $m$ (kg/s) | $p$ (MPa) | $T$ (°C) | $h$ (kJ/kg) | $s$ (kJ/kg.K) |
|---|---|---|---|---|---|---|
| 1 | Ethane | 0.57 | 0.10 | −88.82 | −0.60 | 0.00 |
| 2 | Ethane | 0.57 | 2.82 | −87.08 | 6.54 | 0.01 |
| 3 | Ethane | 0.57 | 2.82 | 15.00 | 573.44 | 2.22 |
| 4 | Ethane | 0.57 | 0.10 | −88.82 | 442.42 | 2.65 |
| 5 | Ammonia | 0.28 | 0.05 | −46.52 | 133.65 | 0.63 |
| 6 | Ammonia | 0.28 | 0.70 | −46.34 | 134.98 | 0.64 |
| 7 | Ammonia | 0.28 | 0.70 | 15.00 | 1622.02 | 5.93 |
| 8 | Ammonia | 0.28 | 0.05 | −46.52 | 1391.99 | 6.84 |
| 9 | LNG | 1.00 | 0.12 | −162.00 | −1.77 | −0.02 |
| 10 | LNG | 1.00 | 6.00 | −158.77 | 18.01 | 0.04 |
| 11 | LNG | 1.00 | 6.00 | −93.82 | 268.40 | 1.75 |
| 12 | LNG | 1.00 | 6.00 | −51.52 | 616.16 | 3.49 |
| 13 | LNG | 1.00 | 6.00 | 15.00 | 823.63 | 4.32 |
| 14 | LNG | 1.00 | 0.70 | −90.32 | 648.14 | 4.57 |
| 15 | LNG | 1.00 | 0.70 | 10.00 | 870.27 | 5.54 |
| Overall | CRR | 32.82% | | | | |

A cold energy storage-based power generation system is designed to mitigate the capacity limitation and energy mismatch caused by the varying LNG regasification rate [43]. The schematic diagram of the system is illustrated in Figure 4.26a. In this system, the LNG cold energy is not directly sent to drive the Rankine cycle. Instead, a cold energy storage unit is adopted to first store the LNG cold energy. Then, the stored cold energy is employed as the heat sink to drive the Rankine cycle as and when it is required. For instance, the LNG cold energy can be stored during the nighttime to minimize power generation. On the other hand, the stored cold energy is gradually released to maximize power generation during peak times. Meanwhile, the capacity of power generation units for stable power output will not be limited by the minimum LNG regasification rate.

The waste heat from adjacent power plants can be integrated and linked up to enhance the turbine inlet parameters of the Rankine cycle. A growing number of LNG receiving terminals supply regasified natural gas for an adjacent natural gas-driven power plant, providing an anchor market for the LNG receiving facility [44]. Therefore, using the waste heat from adjacent power plants should be a potentially viable option. Meanwhile, a non-toxic, non-flammable working medium, such as $CO_2$, can be chosen for the Rankine cycle to improve safety and eco-friendliness. The T-Q diagram of the cold energy storage-based power generation system is depicted in Figure 4.26b.

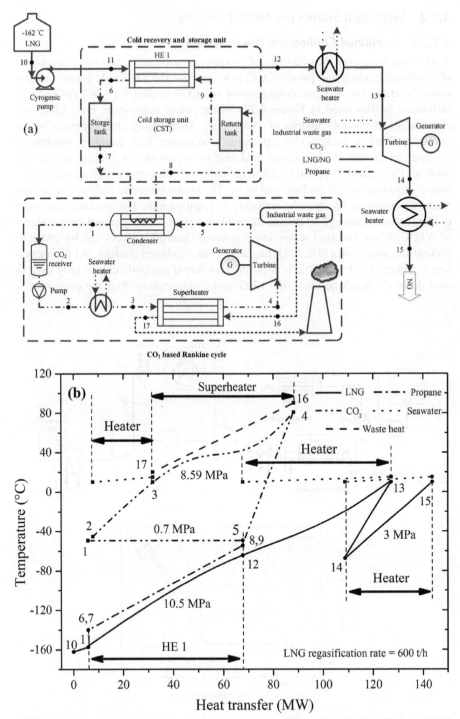

**FIGURE 4.26** Cold energy storage-based power generation system: (a) schematic diagram, (b) T-Q diagram.

## 4.2.2 INTEGRATED SYSTEMS FOR MULTI-PRODUCTS

### 4.2.2.1 Combined Cooling and Power System

Aside from increasing the efficiency of power generation units, one other key aspect of combined cooling and power (CCP) is to improve the LNG cold energy utilization efficiency. The schematic diagram of the CCP system for LNG cold energy utilization is illustrated in Figure 4.27a. After power generation, most LNG still contains a considerable amount of cold energy. Additionally, the evaporator of the cryogenic Rankine cycle is also a potential cold source. It is, therefore, intuitive to continue to harness the remaining cold energy to provide for cooling needs. The cooling effects can be applied to different fields, including space cooling, cold warehouse, compressor inlet cooling, and so on. The medium used to deliver cold energy is chosen based on the specific temperature requirements. For instance, for space cooling, water is enough to deliver the cold energy at delivery/return temperatures of 7/12°C. If the required temperature is lower than 0°C, glycol can be added to prevent freezing issues. The T-Q diagram of the combined cooling and power system is depicted in Figure 4.27b. It is apparent that if the LNG cold is only used to cool down the Rankine cycle, the LNG outlet temperature after the condenser is

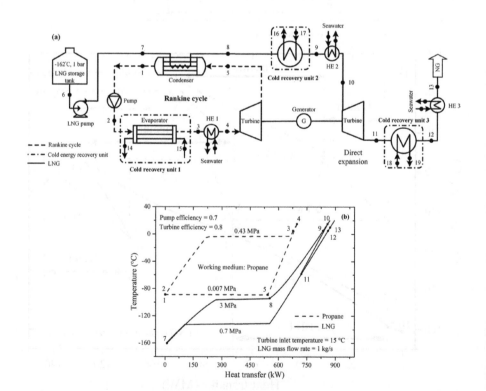

**FIGURE 4.27** Combined cooling and power system for LNG cold energy utilization: (a) schematic diagram; (b) T-Q diagram.

still low. If the cold energy can be further recovered instead of being discharged to the seawater, the cold energy utilization efficiency can be improved significantly. Correspondingly, the cold energy recovery rate can be expressed as

$$\text{CRR} = \frac{W_{net} + Q_c}{Q_{lng}} = \frac{W_{net} + Q_c}{m_{lng}(h_{13} - h_6)} \tag{4.62}$$

The net power output and cooling production are respectively defined as

$$W_{net} = W_{t,\text{Rankine}} + W_{t,lng} - W_{p,\text{Rankine}} - W_{p,lng} \tag{4.63}$$

$$Q_c = m_{wm}(h_3 - h_2) + m_{lng}(h_9 - h_8) + m_{lng}(h_{12} - h_{11}) \tag{4.64}$$

Table 4.9 summarizes the detailed thermodynamic parameters for the combined cooling and power system. Propane is still used as the working medium for the Rankine cycle. The condensing temperature of the Rankine cycle is designed to close the LNG evaporating temperature as much as possible, i.e., a minimum 5°C temperature difference. When the natural gas leaves the condenser, the temperature is still −93.9°C. Comparatively, the propane temperature leaving the pumps in the Rankine

**TABLE 4.9**
**Detailed Thermodynamic Parameters for Combined Cooling and Power Cycle**

| Points | Fluid | m (kg/s) | p (MPa) | T (°C) | h (kJ/kg) | s (kJ/kg.K) |
|---|---|---|---|---|---|---|
| 1 | Propane | 1.16 | 0.007 | −88.90 | −0.60 | 0.12 |
| 2 | Propane | 1.16 | 0.43 | −88.66 | 0.36 | 0.12 |
| 3 | Propane | 1.16 | 0.43 | 5.00 | 585.38 | 2.43 |
| 4 | Propane | 1.16 | 0.43 | 15.00 | 602.71 | 2.49 |
| 5 | Propane | 1.16 | 0.007 | −88.90 | 469.15 | 2.67 |
| 6 | LNG | 1.00 | 0.12 | −162.00 | −1.77 | −0.02 |
| 7 | LNG | 1.00 | 3.00 | −160.41 | 7.93 | 0.01 |
| 8 | LNG | 1.00 | 3.00 | −93.90 | 551.22 | 3.40 |
| 9 | LNG | 1.00 | 3.00 | 5.00 | 832.70 | 4.68 |
| 10 | LNG | 1.00 | 3.00 | 15.00 | 856.91 | 4.76 |
| 11 | LNG | 1.00 | 0.70 | −58.42 | 719.16 | 4.93 |
| 12 | LNG | 1.00 | 0.70 | 5.00 | 859.09 | 5.50 |
| 13 | LNG | 1.00 | 0.70 | 10.00 | 870.28 | 5.54 |
| 14 | Water | 32.25 | 0.10 | 12.00 | 50.50 | 0.18 |
| 15 | Water | 32.25 | 0.10 | 7.00 | 29.52 | 0.11 |
| 16 | Water | 13.42 | 0.10 | 12.00 | 50.50 | 0.18 |
| 17 | Water | 13.42 | 0.10 | 7.00 | 29.52 | 0.11 |
| 18 | Water | 6.67 | 0.10 | 12.00 | 50.50 | 0.18 |
| 19 | Water | 6.67 | 0.10 | 7.00 | 29.52 | 0.11 |
| Overall | CRR | 157.8% | | | | |

**TABLE 4.10**

**The Enthalpy Changes and Work Output/Consumption for Combined Cooling and Power Cycle**

| Components/Working Mediums | $\Delta h$ (kW) | W (kW) |
|---|---|---|
| LNG | 872.05 | – |
| Cold recovery unit 1 | −676.62 | – |
| Cold recovery unit 2 | −281.48 | – |
| Cold recovery unit 3 | −139.93 | – |
| HE 1 | −20.05 | – |
| HE 2 | −24.21 | – |
| HE 3 | −11.18 | – |
| Work output of direct expansion turbine | – | 135.00 |
| Work output of Rankine cycle turbine | – | 151.39 |
| Mechanical losses | – | 5.84 |
| LNG pump work consumption | – | −1.11 |
| Rankine cycle pump work consumption | – | −9.71 |
| $W + \Delta h$ | 0 | |

cycle is −88.6°C. The natural gas after the direct expansion turbine is −58.4°C. All these cold energies can, collectively, be converted to useful cooling effects for space cooling or other purposes. Finally, if the cold energy is converted to chilled water at 7°C, the cold energy recovery rate approaches 157.8%. It is worth noting that it does not mean this system violates energy conservation law. The key step to produce the extra cooling effect is process 10–11, where the physical exergy of natural gas is converted to work output. During the process, adiabatic expansion makes the natural gas temperature decrease again. From the whole process, the reason why the cold energy recovery rate is larger than 100% is that high-grade cold energy is converted to low-grade cold energy. In other words, the LNG temperature is about −160°C, whose energy quality is much higher than the chilled water at 7°C. The principle is similar to the air conditioner, and the COP is usually larger than 1 (specifically 3–5 for domestic use), where the high-grade power energy is converted to low-grade cooling effects. Therefore, the efficiency calculated from the first law of thermodynamics does not consider the energy quality. If the efficiency is calculated from the exergy perspective, the efficiency must be smaller than 1. The detailed calculations based on the first law of thermodynamics are listed in Table 4.10. Assuming that the components are adiabatic, the result shows that the system strictly follows the first law of thermodynamics ($W + \Delta h = 0$).

### 4.2.2.2 Cold Energy Storage-Based Multigeneration Systems

Despite many multigeneration systems having reported higher efficiency, most of them do not consider the demand needs of the downstream users. In other words, the performance of these systems is only evaluated from the system perspective. Energy mismatch often occurs if the LNG regasification rate is not judiciously synchronized with users' demands. For instance, the ratio of power to cooling is fixed at the design condition. Under off-design conditions, the power and cooling productions are often interactively dynamic and cannot be adjusted independently. Accordingly, the energy mismatch between cold recovery systems and users is the underlying problem that needs to be addressed.

To resolve this issue, a cold energy storage-based multigeneration system can be designed and employed. The schematic diagram of the system is shown in Figure 4.28a. There are two regasification lines in the system. The main line is the cold energy recovery line, where the LNG cold energy is converted to utilities. To

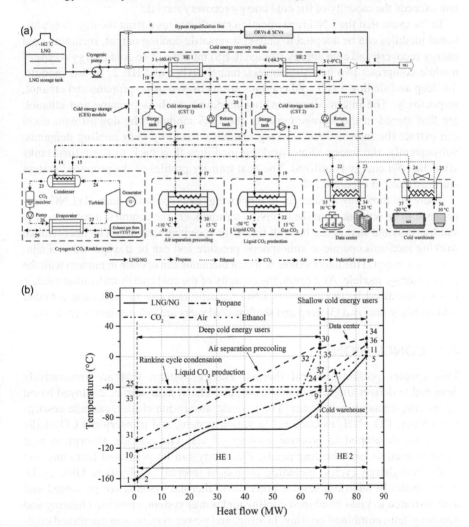

**FIGURE 4.28** (a) Heat transfer curves of the proposed system (LNG regasification rate = 367.2 t/h); (b) schematic diagram of a cold energy storage-based multigeneration system for LNG cold energy utilization [46].

improve the utilization efficiency, the LNG cold energy is utilized in a cascading way. They are divided into deep and shallow cold energies. The deep cold can be employed for cryogenic power generation, air separation precooling, and liquid $CO_2$ production while the possible shallow cold energy provides the cooling needs of storage warehouses and data centers [45].

There is another bypass regasification line, where either open rack vaporizers (ORVs) or submerged combustion vaporizers (SCVs) are installed. The bypass line is employed to ensure the safe operation of the LNG receiving terminal in the event that a shutdown of the cold energy recovery module happens. In addition, the bypass regasification line can vaporize the surplus LNG when the required regasification rate exceeds the capacity of the cold energy recovery module.

In the event that the LNG regasification rate fluctuates during the day, two additional modules can be adopted to deliver a versatile cooling output, including cold energy recovery and cold energy storage (CES) modules. The cold energy recovery module comprises two heat exchangers, namely HE 1 and HE 2. In this module, the deep and shallow LNG cold energies are recovered using propane and ethanol, respectively. The intermediate working mediums, such as propane and ethanol, are first stored in the storage tanks in the CES module. The downstream users can extract the intermediate working mediums based on their cooling demands. Subsequently, the intermediate working mediums flow back to the return tanks after the cold energy is utilized. The heat transfer profiles of the proposed system are illustrated in Figure 4.28b.

There are three key advantages of employing the proposed system: (1) LNG does not need to be sent out from the receiving terminal. On the contrary, a cold recovery and transmission station can be built inside the receiving terminal. The intermediate working mediums operate at atmospheric pressure and can be transported in a relatively easy way; (2) the intermediate working mediums can operate in tandem with the energy storage module. As a result, the capacity of the cold energy utilization module is not constrained by the minimum LNG regasification rate to implement a higher cold recovery rate; and (3) deep and shallow cold users are independent of each other.

## 4.3 CONCLUSION

This chapter presents several thermal energy systems that are innovatively designed to deliver single or multiple outputs. The systems can be deployed based on specific demands and needs. The cascade absorption chiller, cascade adsorption chiller, $CO_2$–$NH_3$ cascade refrigeration system, and transcritical $CO_2$–LiBr cycles are developed to produce cooling effects. The cascade absorption heat pump is used to meet heating needs. Electricity can be generated from any one of the heat-driven cycles, including combined internal combustion, ORC cycle, or KC cycle. In addition, several other novel thermal systems are presented and demonstrated to yield combined cooling and power system, combined heating and power system, combined cooling, heating, and power system, and combined cooling, heating, potable water, and power systems. Lastly, various integrated systems are also proposed to illustrate how thermal energy from cold energy can be harnessed to produce useful outputs.

# REFERENCES

[1] Cui P, Yu M, Liu Z, Zhu Z, Yang S. Energy, exergy, and economic (3E) analyses and multi-objective optimization of a cascade absorption refrigeration system for low-grade waste heat recovery. *Energy Conversion and Management* 2019;184:249–61. https://doi.org/10.1016/j.enconman.2019.01.047

[2] Alberto Dopazo J, Fernández-Seara J, Sieres J, Uhía FJ. Theoretical analysis of a $CO_2$–$NH_3$ cascade refrigeration system for cooling applications at low temperatures. *Applied Thermal Engineering* 2009; 29:1577–83. https://doi.org/10.1016/j.applthermaleng.2008.07.006

[3] Dakkama HJ, Elsayed A, AL-Dadah RK, Mahmoud SM, Youssef P. Integrated evaporator–condenser cascaded adsorption system for low temperature cooling using different working pairs. *Applied Energy* 2017;185:2117–26. https://doi.org/10.1016/j.apenergy.2016.01.132

[4] Su B, Han W, Sui J, Jin H. A two-stage liquid desiccant dehumidification system by the cascade utilization of low-temperature heat for industrial applications. *Applied Energy* 2017;207:643–53. https://doi.org/10.1016/j.apenergy.2017.05.184

[5] Xu ZY, Gao JT, Mao HC, Liu DS, Wang RZ. Double-section absorption heat pump for the deep recovery of low-grade waste heat. *Energy Conversion and Management* 2020;220:113072. https://doi.org/10.1016/j.enconman.2020.113072

[6] He M, Zhang X, Zeng K, Gao K. A combined thermodynamic cycle used for waste heat recovery of internal combustion engine. *Energy* 2011;36:6821–9. https://doi.org/10.1016/j.energy.2011.10.014

[7] Yang S, Deng C, Liu Z. Optimal design and analysis of a cascade LiBr/$H_2O$ absorption refrigeration/transcritical $CO_2$ process for low-grade waste heat recovery. *Energy Conversion and Management* 2019;192:232–42. https://doi.org/10.1016/j.enconman.2019.04.045

[8] Wang J, Liu Z, Wang H, Liu X. A new combined cooling and power system based on ammonia-water absorption refrigeration cycle: Thermodynamic comparison and analysis. *Energy Conversion and Management* 2022;270:116262. https://doi.org/10.1016/j.enconman.2022.116262

[9] Zoghi M, Habibi H, Chitsaz A, Javaherdeh K, Ayazpour M. Exergoeconomic analysis of a novel trigeneration system based on organic quadrilateral cycle integrated with cascade absorption-compression system for waste heat recovery. *Energy Conversion and Management* 2019;198:111818. https://doi.org/10.1016/j.enconman.2019.111818

[10] Chen WD, Chua KJ. Energy, exergy, economic, and environment (4E) assessment of a temperature cascading multigeneration system under experimental off-design conditions. *Energy Conversion and Management* 2022;253:115177. https://doi.org/10.1016/j.enconman.2021.115177

[11] Jiang R, Han W, Qin FGF, Sui J, Yin H, Yang M. Thermodynamic model development, experimental validation and performance analysis of a MW CCHP system integrated with dehumidification system. *Energy Conversion and Management* 2018;158:176–85. https://doi.org/10.1016/j.enconman.2017.12.060

[12] Han W, Chen Q, Lin R, Jin H. Assessment of off-design performance of a small-scale combined cooling and power system using an alternative operating strategy for gas turbine. *Applied Energy* 2015;138:160–8. https://doi.org/10.1016/j.apenergy.2014.10.054

[13] Afzali SF, Mahalec V. Optimal design, operation and analytical criteria for determining optimal operating modes of a CCHP with fired HRSG, boiler, electric chiller and absorption chiller. *Energy (Oxford)* 2017;139:1052–65. https://doi.org/10.1016/j.energy.2017.08.029

[14] Al-Karaghouli A, Kazmerski LL. Energy consumption and water production cost of conventional and renewable-energy-powered desalination processes. *Renewable & Sustainable Energy Reviews* 2013;24:343–56. https://doi.org/10.1016/j.rser.2012.12.064

[15] Moghimi M, Emadi M, Ahmadi P, Moghadasi H. 4E analysis and multi-objective optimization of a CCHP cycle based on gas turbine and ejector refrigeration. *Applied Thermal Engineering* 2018;141:516–30, https://doi.org/10.1016/j.applthermaleng.2018.05.075

[16] Anvari S, Mahian O, Taghavifar H, Wongwises S, Desideri U. 4E analysis of a modified multigeneration system designed for power, heating/cooling, and water desalination. *Applied Energy* 2020;270:115107. https://doi.org/10.1016/j.apenergy.2020.115107

[17] Köse Ö, Koç Y, Yağlı H. Energy, exergy, economy and environmental (4E) analysis and optimization of single, dual and triple configurations of the power systems: Rankine cycle/Kalina cycle, driven by a gas turbine. *Energy Conversion and Management* 2021;227:113604. https://doi.org/10.1016/j.enconman.2020.113604

[18] Wang J, Lu Z, Li M, Lior N, Li W. Energy, exergy, exergoeconomic and environmental (4E) analysis of a distributed generation solar-assisted CCHP (combined cooling, heating and power) gas turbine system. *Energy* 2019;175:1246–58. https://doi.org/10.1016/j.energy.2019.03.147

[19] Bejan A, Tsatsaronis G, Moran MJ. *Thermal design and optimization.* New York: John Wiley; 1996.

[20] Dinçer İ, Rosen M. *Exergy: Energy, environment, and sustainable development.* Amsterdam; Boston: Elsevier; 2007.

[21] Zhang Y, Li B, Li H, Zhang B. Exergy analysis of biomass utilization via steam gasification and partial oxidation. *Thermochimica Acta* 2012;538:21–8. https://doi.org/10.1016/j.tca.2012.03.013

[22] Avanessian T, Ameri M. Energy, exergy, and economic analysis of single and double effect LiBr–H$_2$O absorption chillers. *Energy and Buildings* 2014;73:26–36. https://doi.org/10.1016/j.enbuild.2014.01.013

[23] Anand S, Gupta A, Tyagi SK, Anand Y. An absorption chiller system using lithium bromide and water as working fluids: Energy analysis. *ASHRAE Transactions* 2014;120:226.

[24] Alsarayreh AA, Al-Maaitah A, Attarakih M, Bart H-J. Energy and exergy analyses of adsorption chiller at various recooling-water and dead-state temperatures. *Energies* 2021;14. https://doi.org/10.3390/en14082172

[25] Thu K, Saha BB, Chua KJ, Bui TD. Thermodynamic analysis on the part-load performance of a microturbine system for micro/mini-CHP applications. *Applied Energy* 2016;178:600–8. https://doi.org/10.1016/j.apenergy.2016.06.106

[26] Kong XQ, Wang RZ, Huang XH. Energy efficiency and economic feasibility of CCHP driven by stirling engine. *Energy Conversion and Management* 2004;45:1433–42. https://doi.org/10.1016/j.enconman.2003.09.009

[27] Hou Q, Zhao H, Yang X. Economic performance study of the integrated MR-SOFC-CCHP system. *Energy* 2019;166:236–45. https://doi.org/10.1016/j.energy.2018.10.072

[28] Al-Karaghouli A, Kazmerski L. Economic and technical analysis of a reverse-osmosis water desalination plant using DEEP-3.2 software. *Journal of Environmental Science Engineering* 2012;1. https://www.davidpublisher.com/index.php/Home/Article/index?id=5139.html

[29] Chen J, Li X, Dai Y, Wang C-H. Energetic, economic, and environmental assessment of a Stirling engine based gasification CCHP system. *Applied Energy* 2021;281:116067. https://doi.org/10.1016/j.apenergy.2020.116067

[30] Wang S, Fu Z. Thermodynamic and economic analysis of solar assisted CCHP-ORC system with DME as fuel. *Energy Conversion and Management* 2019;186:535–45. https://doi.org/10.1016/j.enconman.2019.02.076

[31] Chen Q, Ja MK, Li Y, Chua KJ. Energy, economic and environmental (3E) analysis and multi-objective optimization of a spray-assisted low-temperature desalination system. *Energy* 2018;151:387–401. https://doi.org/10.1016/j.energy.2018.03.051

[32] Li R, Dai Y, Cui G. Multi-objective optimization of solar powered adsorption chiller combined with river water heat pump system for air conditioning and space heating application. *Energy* 2019;189:116141. https://doi.org/10.1016/j.energy.2019.116141

[33] Pierobon L, Nguyen T-V, Larsen U, Haglind F, Elmegaard B. Multi-objective optimization of organic Rankine cycles for waste heat recovery: Application in an offshore platform. *Energy* 2013;58:538–49. https://doi.org/10.1016/j.energy.2013.05.039

[34] Wang L, Lu J, Wang W, Ding J. Energy, environmental and economic evaluation of the CCHP systems for a remote island in south of China. *Applied Energy* 2016;183:874–83. https://doi.org/10.1016/j.apenergy.2016.09.023

[35] Hou H, Wu J, Ding Z, Yang B, Hu E. Performance analysis of a solar-assisted combined cooling, heating and power system with an improved operation strategy. *Energy* 2021;227:120516. https://doi.org/10.1016/j.energy.2021.120516

[36] Arens M, Åhman M, Vogl V. Which countries are prepared to green their coal-based steel industry with electricity? Reviewing climate and energy policy as well as the implementation of renewable electricity. *Renewable and Sustainable Energy Reviews* 2021;143:110938. https://doi.org/10.1016/j.rser.2021.110938

[37] Neven TV. *Fuel, electricity, and natural gas price*. GlobalPetrolPrices.com. API; 2021.

[38] Xu XY, Ang BW. Analysing residential energy consumption using index decomposition analysis. *Applied Energy* 2014;113:342–51. https://doi.org/10.1016/j.apenergy.2013.07.052

[39] Energy Market Authority. Average monthly household electricity consumption by dwelling type. 2020. https://data.gov.sg/dataset/average-monthly-household-electricity-consumption-by-dwelling-type

[40] Energy Market Authority. *Electricity and town gas tariffs*. Singapore: Government of Singapore; 2020.

[41] Kanagawa T. *Japan's LNG utilization and environmental efforts*. Tokyo, Japan: The Japan Gas Association; 2008.

[42] Kim KST. LNG cold energy utilization technology. In: Zhang X., Dincer I, editors. *Energy solutions to combat global warming*. vol. 33. Midtown Manhattan, NY: Springer; 2017.

[43] Huang ZF, Wan YD, Soh KY, Islam MR, Chua KJ. Off-design and flexibility analyses of combined cooling and power based liquified natural gas (LNG) cold energy utilization system under fluctuating regasification rates. *Applied Energy* 2022;310:118529. https://doi.org/10.1016/j.apenergy.2022.118529

[44] Stradioto DA, Seelig MF, Schneider PS. Performance analysis of a CCGT power plant integrated to a LNG regasification process. *Journal of Natural Gas Science and Engineering* 2015;27:18–22. https://doi.org/10.1016/j.jngse.2015.06.009

[45] Jahangir MH, Mokhtari R, Mousavi SA. Performance evaluation and financial analysis of applying hybrid renewable systems in cooling unit of data centers – A case study. *Sustainable Energy Technologies and Assessments* 2021;46:101220. https://doi.org/10.1016/j.seta.2021.101220

[46] Huang ZF, Soh KY, Wan YD, Islam MR, Chua KJ. Assessment of an intermediate working medium and cold energy storage (IWM-CES) system for LNG cold energy utilization under real regasification case. *Energy* 2022;253:124080. https://doi.org/10.1016/j.energy.2022.124080

# 5 Challenges and Prospects for Waste Thermal Energy Recovery

## 5.1 CHALLENGES AND PROSPECTS FOR HEAT ENERGY RECOVERY

### 5.1.1 FINANCIAL SUPPORT AND GOVERNMENT POLICY

Albeit acknowledging the importance of recovering thermal energy, it is still being reported that a considerable amount of waste heat is continuously being dissipated directly into the ambient [1]. One potential reason is that the payback period of deploying a waste heat recovery system is too long. This is attributed to the financial burden laden on the industry as the initial installation investment to recover waste heat is enormous. Further, it is also challenging to realize large-scale waste heat recovery applications without good governmental support policies as well as the establishment of stable carbon trading markets.

Substantial governmental support is crucial to encourage industries to adopt waste thermal energy utilization. For example, the United Kingdom has proposed an Industrial Strategy plan [2] to support thermal energy recovery in industrial activities. A key component of the Industrial Strategy is to deliver affordable energy and clean growth, which recognizes energy productivity as a way to reduce energy costs and improve economic productivity. Industrial heat recovery can improve energy productivity by reducing fuel requirements and minimizing energy waste. With the implementation of the Industrial Strategy Programme, industrial energy bills could potentially be reduced by over £500 million, enhancing competitiveness and protecting the industry from future energy price fluctuations. Energy efficiency can be improved by recovering heat from industrial processes, resulting in significant energy savings. Thus far, data have suggested that a relatively low level of support could significantly impact the deployment of economically viable industrial heat recovery strategies [2].

A 2014 report produced by Element Energy [3] showed that 11 TWh/y of industrial heat use could have been recovered at a technical level from industrial processes in eight key energy-intensive sectors, of which 5 TWh/y would have been commercially viable [3]. It was evident from the stakeholders' perspective that there was a need for significant support to overcome barriers to accessing finance, including investments that did not meet internal hurdle rates and external financing costs that were high, as well as the availability of information barriers that needed to be addressed. In addition, according to a previous government's 2013 Heat Strategy [4], waste heat can be used to decarbonize the industry by recovering and reusing it. A key decarbonization option identified by the Industrial Decarbonization and

Energy Efficiency Roadmaps project [5] involves thermal energy recovery. Under the maximum technical potential for decarbonization in those sectors, industrial heat recovery can save up to 1.75 Mt $CO_2$ by 2050, or 3% of cross-sector carbon reductions. Heat recovery was also recognized as an important element of decarbonizing nonresidential buildings in the fifth carbon budget [6] which was published by the Committee on Climate Change in October 2016.

### 5.1.2　HEAT ENERGY STORAGE AND TRANSPORTATION

Industries, which produce a large amount of waste heat, are sometimes located far away from the consumers. Thus, research on the employment of storage and transport of thermal energy has become a key endeavor in waste heat energy utilization. In addition, even when the waste heat generation site is close to the consumers, a heat thermal energy storage system is often relied on to store recovered thermal energy. This is because consumers' energy demands frequently fluctuate, making it challenging to utilize the waste heat efficiently. Therefore, initial financial investment for thermal heat storage and efficient transport system has led to more significant deployment difficulties for waste heat recovery systems.

Buildings consume approximately 39% of all primary energy and 74% of all electricity in the United States [7]. About 50% of a building's energy demand is derived from thermal energy applications. In order to achieve the target of decarbonized buildings by 2050, the employment of thermal energy storage will be a critical enabler. With recent advancements made in thermal energy-storing technologies, energy savings is expected to increase. Heat pump, a conventional thermal technology, would be more energy-efficient and affordable, and building loads could be significantly shed or shifted more easily, which would lead to improved thermal comfort for occupants. By improving the spatial and temporal control of heat flow, storage capacity can be utilized more efficiently, and system costs can accordingly be reduced. However, an additional heat transfer fluid is usually required to conduct heat exchange between the phase change material and the user side during heat release, which can result in poor energy efficiency due to an additional heat transfer segment.

### 5.1.3　IMPROVEMENT OF WASTE HEAT RECOVERY USING ADVANCED TECHNOLOGIES

Despite many waste heat recovery technologies being developed, experimentally tested, and commercially deployed, significant shortages of high-performing waste heat recovery systems exist.

In the past, research pertaining to improving the energy efficiency of thermal energy processes focused on two chief aspects: experimental study and simulations based on physical models. There is no doubt that these two approaches are important to developing energy-efficient thermal energy processes. However, there are also apparent drawbacks to these methods. Experimental research is a reliable method to unveil the underlying mechanisms of thermal energy processes. However, massive investment often hinders research progress through this path. On the contrary, well-designed simulation models based on governing laws of physics are helpful to discover and understand the physical phenomena behind thermal energy processes. However, they are sometimes effort-intensive and time-consuming to achieve high-resolution results, such as direct numerical simulation in computational fluid dynamics.

With the rapid development of computer science and communication technologies, the idea of integrating thermal energy processes with cutting-edge technologies, such as artificial intelligence (AI), digital twin, blockchain, and metaverse, has become an intriguing platform for research. AI can mimic human reasoning and cognition to learn and solve problems [8]. Further, AI can deal with uncertain and imprecise information, which makes it suitable for handling complex real-life problems [9]. The AI-based models usually require fewer computational resources and do not rely on physical knowledge [10]. Also, AI is very adept at handling nonlinear and concave problems. Since AI technologies usually require a large amount of data to learn patterns and train models, performance becomes poor when there is insufficient available data. Recently, the development of the Internet of Things (IoT) has provided an exciting platform for the development of AI-based models. The IoT aims to connect and exchange data between different physical objects through the internet or other communication networks, which indicates a massive amount of data will be produced [11]. Extensive employment of IoT data will definitely boost the development of AI-based models and find potential improvements for thermal energy processes. Moreover, other digital technologies, such as digital twin, blockchain, and metaverse, have been extensively applied in different fields. Despite the fact that some of these advanced digital technologies have emerged from different fields, for instance, blockchain was first used in cryptocurrency, applying these technologies to the energy sector has fascinated many researchers. In addition, these digital technologies are evolved from computer science and communication, which are naturally coherent and integrable.

Figure 5.1 illustrates the technical integration of the metaverse concept with systems operating in the real world. Evolving a better 5G/6G network is the key supporting platform to realize the metaverse paradigm. The cloud computing section is designed for fast data processing. Blockchain technology provides an authentication mechanism and consensus mechanism. In this way, the metaverse can replace the model of data misappropriation experienced by centralized agencies and fully protects data rights. Based on augmented reality and virtual reality systems, operators can give a command and receive information from the metaverse. Digital twin and artificial intelligence are the key components of the metaverse. Additionally, data and systems from the real world can be precisely represented in the metaverse by employing digital twin and artificial intelligence.

### 5.1.4 Prospects On Waste Heat Recovery

Research and development on advanced materials may contribute to improving the efficiencies of waste heat recovery technologies. For example, the development of more advanced adsorbent materials promotes the efficiencies of adsorption chillers and solid/liquid desiccant dehumidifiers.

Research studies involving better thermal storage materials are also key to achieving a low-carbon future [1]. That is because a better-performing thermal storage technology

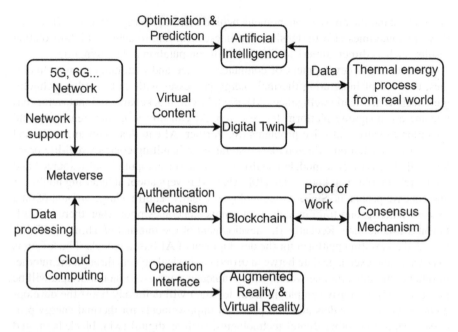

**FIGURE 5.1** Technical diagram of metaverse integrated with thermal energy process systems from the real world.

not only balances well the load matching between supply and demand but also stables the energy supply chain by considering the seasonal peak and trough of demands.

It is anticipated that more advanced waste heat recovery technologies will be designed and engineered. For example, the s-$CO_2$ cycle has demonstrated better thermal efficiency than the conventional cycle [12].

Finally, a mature and stable carbon trading market coupled with supportive governmental policies is expected to promote the research and development of waste heat recovery technologies.

## 5.2 CHALLENGES AND PROSPECTS FOR WASTE COLD ENERGY RECOVERY

### 5.2.1 POLICY AND PLANNING

Due to safety concerns and economic issues, LNG cannot be sent out from terminals for long-distance transportation. Thus, LNG cold energy utilization setups are usually deployed inside or near the LNG receiving terminal, typically within 1 km. LNG receiving terminals are often located in the vicinity of coastal areas, which sometimes belong to remote areas. On the contrary, cold energy users need to import raw materials and export their products, which requires a developed transportation network. Therefore, it is not easy to arrange users around the receiving terminals. Project sites and supporting infrastructures are key factors in setting up a feasible industrial chain for LNG cold energy utilization. In addition, the industrial chain involves different companies and authorities who have their own vested interests, and local government may need to carry out relevant policies and play the role of coordinator to guide the industrial chain.

Some receiving terminals have no plan to recover cold energy during the beginning stage of their designs. These LNG-receiving terminals need to reconsider their cold energy recovery setups to adhere to new governmental regulations. Indeed, the development of an efficient cold energy utilization facility for such a receiving terminal could pose to be an enormous challenge.

## 5.2.2 Low Exergy Efficiency and Overall Recovery Rate Issue

If the LNG cold energy is only used to produce shallow cold energy, such as chilled water (~5°C), then almost all the LNG cold energy can be recovered. However, the inevitable exergy destruction will be significant. This is because the high-quality LNG cold energy (−162°C) is converted to low-quality cold energy. For generating power, the theoretical exergy efficiency spans 12%–30%, according to our case studies in previous chapters. In practice, the exergy efficiency of commissioned plants lingers around 8%–20% [13]. Directly converting the LNG cold energy to the other working mediums with similar low temperatures is deemed to be an efficient method. This is because the cold exergy is retained by other working mediums. Therefore, the efficient system for LNG cold energy utilization ought to involve a cascading system in which the LNG temperature matches with required temperatures well. As shown in Figure 5.2, the deep cold energy may be first applied for the purpose of air

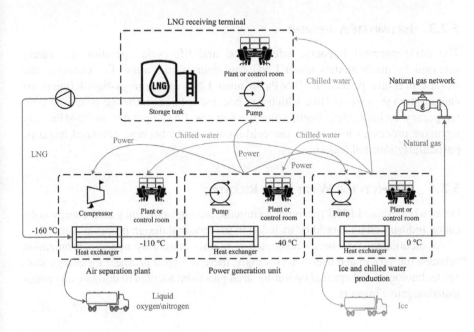

**FIGURE 5.2** Cascading system for LNG cold energy utilization.

separation where the air needs to be cooled down to below −160°C. Then, the natural gas leaving the air separation plant (approximately −110°C) can be used for power generation or dry ice production. Finally, the rest of LNG cold energy is employed for ice-making or chilled water production.

Cascading systems have been proposed by many researchers to improve exergy efficiency. However, most of these studies have merely focused on the performance of the system without considering users' demands. The issue on low-efficiency issue still exists if the energy mismatch between the cold energy recovery system and users' demand is not addressed. In other words, the cold energy users and their demand profiles are varied according to the commissioned LNG cold energy utilization setups [14]. If the cold energy is judiciously deployed in a cascading manner, then the downstream users will be influenced by the upstream users. Therefore, synchronizing the cascading cold energy based on different users needs to be carefully considered.

Fluctuating regasification rates can also result in an energy mismatch that affects the overall cold energy recovery rate. LNG receiving terminals are built to meet the natural gas demand of thermal power plants and city utility needs (including residence and industry). The natural gas demand from these sectors usually appears to have seasonal or hourly variation. However, the cold energy demand may not synchronize with the LNG regasification rate well. Therefore, a case of insufficient cold energy to meet users' needs arises. Satisfying the natural gas demand is the priority for any receiving terminal. Consequently, the capacity of the cold energy utilization installations is usually designed at 20%–30% of their baseload to ensure a reliable cooling supply [15].

### 5.2.3 INCOMPLETE ASSESSMENT

The environmental impacts, safety risks, and life-cycle operation are rarely assessed by many studies about LNG cold energy utilization. For example, the storage pressure peaked at 21 MPa in most LNG cold energy-based liquid air energy storage systems [16], which has become a huge challenge posed by contemporary technologies. Furthermore, when the LNG operates at 30 MPa, any intrusive processes to transfer the cold energy could become potential hazards, particularly internal leakages.

### 5.2.4 PROSPECTS FOR WASTE COLD RECOVERY

For newly designed LNG receiving terminals, the high-efficiency cold energy utilization industry chain should be included in the terminal design proposal plans.

As mentioned, the energy mismatch is mainly caused by non-synchronization between the system and users. Accordingly, there is a great prospect for energy storage technologies and optimal operating strategies to be adopted to mitigate any anticipated negative impacts.

Lastly, a more comprehensive assessment is necessary when a new design is proposed. Besides conducting thermodynamic and economic analyses, the design scope should include environmental impact, safety, and life cycle assessments in order to holistically evaluate the feasibility and reliability of the proposed system.

## 5.3   CONCLUSIONS

This chapter summarized the challenges and prospects for both heat and cold waste thermal energy recovery. Financial and policy support from the government, storage material development, and employment of advanced artificial intelligence technologies have significant effects on the thermal energy recovery and utilization. Policy and planning, incomplete assessment, and low exergy efficiency are the main challenges for cold energy recovery.

## REFERENCES

[1] Miró L, Gasia J, Cabeza LF. Thermal energy storage (TES) for industrial waste heat (IWH) recovery: A review. *Applied Energy* 2016;179:284–301. https://doi.org/10.1016/j.apenergy.2016.06.147

[2] Department for Business, Energy & Industrial Strategy. Industrial Heat Recovery Support Programme, 2017. chrome-extension://efaidnbmnnnibpcajpcglclefindmkaj/https://assets.publishing.service.gov.uk/government/uploads/system/uploads/attachment_data/file/651125/IHRS_Consultation_Document-October_2017.pdf

[3] Paul S, Robert H. The potential for recovering and using surplus heat from industry, 2014. https://www.gov.uk/government/publications/the-potential-for-recovering-and-using-surplus-heat-from-industry

[4] Department of Energy & Climate Change. The future of heating: meeting the challenge, 2013. https://www.gov.uk/government/publications/the-future-of-heating-meeting-the-challenge

[5] Department of Energy & Climate Change, Department for Business, Innovation & Skills. Industrial decarbonisation and energy efficiency roadmaps to 2050, 2015. https://www.gov.uk/government/publications/industrial-decarbonisation-and-energy-efficiency-roadmaps-to-2050

[6] Committee on Climate Change. Next steps for UK heat policy, 2016. chrome-extension://efaidnbmnnnibpcajpcglclefindmkaj/https://www.theccc.org.uk/wp-content/uploads/2016/10/Next-steps-for-UK-heat-policy-Committee-on-Climate-Change-October-2016.pdf

[7] Office of energy efficiency & renewable energy. Thermal energy storage technologies subprogram area, 2020. https://www.energy.gov/eere/buildings/thermal-energy-storage

[8] Russel S, Norvig P. Artificial intelligence—a modern approach 3rd Edition. *The Knowledge Engineering Review* 2012;11:78–9.

[9] Ibrahim D. An overview of soft computing. *Procedia Computer Science* 2016;102:34–8. https://doi.org/10.1016/j.procs.2016.09.366

[10] He Z, Guo W, Zhang P. Performance prediction, optimal design and operational control of thermal energy storage using artificial intelligence methods. *Renewable and Sustainable Energy Reviews* 2022;156:111977. https://doi.org/10.1016/j.rser.2021.111977

[11] Li J, Herdem MS, Nathwani J, Wen JZ. Methods and applications for artificial intelligence, big data, internet of things, and blockchain in smart energy management. *Energy and AI* 2023;11:100208. https://doi.org/10.1016/j.egyai.2022.100208

[12] Liu L, Yang Q, Cui G. Supercritical carbon dioxide(s-$CO_2$) power cycle for waste heat recovery: A review from thermodynamic perspective. *Processes* 2020;8(11). https://doi.org/10.3390/pr8111461

[13] Kanagawa T. *Japan's LNG utilization and environmental efforts*. Tokyo, Japan: The Japan gas association; 2008.

[14] Yamamoto T, Fujiwara Y. *The accomplishment of 100% utilization of LNG cold energy*. Kuala Lumpur: World Gas Conference; 2012.

[15] Sung T, Kim KC. LNG cold energy utilization technology. In: Zhang X, Dincer I, editors. *Energy solutions to combat global warming*, Cham: Springer International Publishing; 2017, p. 47–66. https://doi.org/10.1007/978-3-319-26950-4_3

[16] He T, Lv H, Shao Z, Zhang J, Xing X, Ma H. Cascade utilization of LNG cold energy by integrating cryogenic energy storage, organic Rankine cycle and direct cooling. *Applied Energy* 2020;277:115570. https://doi.org/10.1016/j.apenergy.2020.115570

# Index

Note: *Italic* page numbers refer to figures.

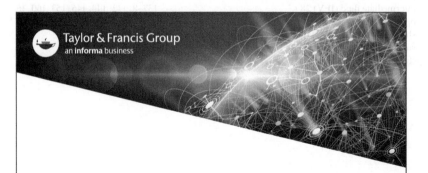

For Product Safety Concerns and Information please contact our
EU representative GPSR@taylorandfrancis.com Taylor & Francis
Verlag GmbH, Kaufingerstraße 24, 80331 München, Germany